AF545181

Numerical Methods & Models in Earth Science

nipa
Browse Subject
Agriculture Sciences (679)
Animal Husbandry and Veterinary Sciences (261)
Climate Change and Environmental Sciences (63)
Community Science (Home Science) (39)
Competitive Examinations and MCQ's (57)
Dairy Technology (91)
Fisheries Sciences (26)
Food Science and Technology (155)
Geological Sciences (30)
Horticultural Sciences (342)
Plant Sciences (191)
Skill and Entrepreneurship Development (47)
eBooks / eChapters / Articles
eBooks
Explore Now
An e book is an electronic book available in various digital formats, can be accessed from anywhere, anytime on desktops/laptops/smart phones and tablets connected to internet.
Explore Now
eBooks / eChapters / Articles
eBooks
Browse, Search, Read & Buy...
Subject Catalogues
eChapters
Publishers
Print Books
Forthcoming
eBooks
New Titles
scan for catalogue
NIPA
GENX
ONLINE RESOURCES
Publishing Books and Journals
Ebooks and e articles, Current Affairs
Reasoning, Logic and Aptitude
Competitive Examination Preparation
Language Learning and Development Programme
Document Quality Checker and Improvement Tool
Effective Public Speaking, Presentation and Interpersonal Skills
Online Programmes for Professional Development
Personality Development and Human Values
PAY USING
UPI
PayPal

Numerical Methods & Models in Earth Science

Editor

Parthasarathi Ghosh
Associate Professor
Geological Studies Unit, Indian Statistical Institute
203, B.T. Road, Kolkata-700 108, West Bengal, India

Indian Statistical Institute
Kolkata, West Bengal, India

2011

New India Publishing Agency
Pitam Pura, New Delhi-110 088

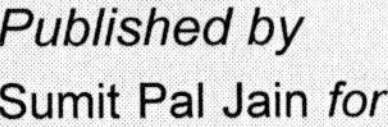

Published by
Sumit Pal Jain *for*

New India Publishing Agency

101, Vikas Surya Plaza, CU Block, L.S.C. Mkt.,
Pitam Pura, New Delhi- 110 088, (India)
Phone: 011-27341717, Fax: 011-27341616
Mobile : 09717133558
E-mail: newindiapublishingagency@gmail.com
Web: www.bookfactoryindia.com

ISBN : 978-93-80235-41-7

Typeset at: Pushpa Bansal *for* Laxmi Art Creations
Printed at: Jai Bharat Printing Press, Delhi

FOREWORD

The Indian Statistical Institute (ISI) was established on 17th December, 1931 by a great visionary Prof. Prasanta Chandra Mahalanobis to promote research in the theory and applications of statistics as a new scientific discipline in India. In 1959, Pandit Jawaharlal Nehru, the then Prime Minister of India introduced the ISI Act in the parliament and designated it as an Institution of National Importance because of its remarkable achievements in statistical work as well as its contribution to economic planning.

Today, the Indian Statistical Institute occupies a prestigious position in the academic firmament. It has been a haven for bright and talented academics working in a number of disciplines. Its research faculty has done India proud in the arenas of Statistics, Mathematics, Economics and Computer Sciences among others. Over seventy five years, it has grown into a massive banyan tree, like the institute emblem. The Institute now serves the nation as a unified and monolithic organization from different places, namely Kolkata, the Headquarters, Delhi, Bangalore, and Chennai, three centers, a network of five SQC-OR Units located at Mumbai, Pune, Baroda, Hyderabad and Coimbatore, and a branch (field station) at Giridih.

The platinum jubilee celebrations of ISI have been launched by Honourable Prime Minister, Prof. Manmohan Singh, on December 24, 2006, and the Govt. of India has declared 29th June as the "Statistics Day" to commemorate the birthday of Prof. Mahalanobis nationally.

Prof. Mahalanobis was a great believer in interdisciplinary research because he thought that this will promote the development not only of Statistics, but also the other natural and social sciences. To promote interdisciplinary research, major strides were made in the areas of computer science, statistical quality control, economics, biological and social sciences, physical and earth sciences.

The Institute's motto of 'unity in diversity' has been the guiding principle in all its activities since its inception. It highlights the unifying role of statistics in relation to various scientific activities.

In tune with this hallowed tradition, a comprehensive academic program, involving Nobel Laureates, Fellows of the Royal Society, Abel prize winner and other dignitaries, has been implemented throughout the Platinum Jubilee year, highlighting the emerging areas of ongoing frontline research in its various scientific divisions, centers, and outlying units. It includes international and national-level seminars, symposia, conferences and workshops, as well as series of special lectures. As an outcome of these events, the Institute is bringing out a series of comprehensive volumes in different subjects including those published under the title *Statistical Science and Interdisciplinary Research* of the World Scientific Press, Singapore.

The present volume titled "Numerical Methods and Models in Earth Science" is one such outcome published by the New India Publishing Agency, New Delhi. It has seven chapters, written by eminent geoscientists, which deal with the modern quantitative methods and numerical models being used to interpret the earth in more objective and meaningful way. The topics include – understanding crustal deformation using geographical positioning system, understanding fluvial transport mechanism with statistical analysis of river sand, study of bone geometry of archaic species, and identifying zones of mineralization using earth science data and GIS techniques. I believe the state-of-the art studies presented in this book through a glimpse of new methodologies will be very useful to researchers as well as practitioners in interpreting earth science data.

Thanks to the contributors for their excellent research contributions, and to the volume editor Dr. Parthasarathi Ghosh for his sincere effort in bringing out the volume nicely. Thanks are also due to New India Publishing Agency, New Delhi for their initiative in publishing the book and being a part of the Platinum Jubilee endeavor of the Institute.

January 2010
Kolkata

SANKAR K. PAL
Director
Indian Statistical Institute
Kolkata

PREFACE

The last couple of decades have witnessed a tremendous research initiative in developing new quantitative methods and models for earth science. The advent of powerful computers equipped with sophisticated mathematical and statistical tools has inspired geo-scientists to attempt more objective interpretation of their data. The earth science data are typically either too numerous or too sparse and are rarely complete. It is thus often difficult to identify a definitive pattern in them or to arrive at a comprehensive abstraction. The recent availability of mathematical and statistical tools, customized for such data sets, provides the opportunity for more objective analyses. Active research is presently underway to devise newer methodologies for analyzing data that were traditionally considered unsuitable for quantitative treatment. The geometry of geological features, that often provide important clues to their formative process, is one such example. Recent advances in the field of Geographical Information Science are now providing a bounty of potential tools for objective analysis of geometry and spatial interrelationship.

With the help of new technologies, it is now possible to study the earth processes that were beyond the scope of observation due problems with their rate, scale or accessibility. Where direct observation is yet not feasible, sophisticated numerical models for explaining geological phenomena are now made increasingly available. By embedding our knowledge of the geological processes in a logical framework, these models allow us to explore areas where our understanding of these processes are flawed or weak, towards which new research efforts are to be directed. On the other hand, a holistic approach to understand earth requires integrating observations from many different sources.

The new technologies have helped this process by providing the facility of sharing earth science data over intranet and Internet. This in turn necessitated development of exclusive standards, protocols and strategies to manage collaborative exchange of data.

It is difficult to encompass this whole wave of new developments in a single book. However, the seven original research papers presented here provide the reader with a glimpse of new quantitative methods and numerical models that are presently used to interpret the earth in a more meaningful way. In the first two chapters Mukul and his co-workers and Mallick and her co-workers demonstrate how the geographic positioning system, that can measure plate motions with unprecedented accuracy, may be used in understanding crustal deformations in eastern Himalayas. The detail of a numerical model to study the relationship between crustal shortening and the resulting rock fabrics, developed by Mitra and his co-authors, is given in the third chapter. In the fourth chapter Purkait shows how statistical analysis of modern river sand helps in understanding fluvial transport mechanisms. Dutta and his co-workers propose a new quantitative technique for comparative study of bone geometry of archaic species in chapter five. In the sixth chapter Porwal and his co-authors demonstrate how zones of mineralization can be identified from earth science data sets with the help of GIS techniques. In the seventh chapter, Saha and his co-authors address the methodologies and strategies adopted for enterprise level earth science data collaboration and management tasks.

I would like to thank F. P. Agterberg (GSC, Canada), A. Aoudia (ASICTP, Trieste Italy), P. Banerjee (WIHG, Dehradun, India), P. Bourke (UWA, Crawley, Australia), S. Chatterjee (Minnesota), P. Dasgupta (Presidency College, Kolkata, India), B. L. Deekshatulu (Hyderabad), J. Ghosh (Purdue), R. W. King (MIT), B. Köbben (ITC, The Netherlands), G. Mitra (Rochester), D. K. Mukhopadhyay (IIT, Roorke, India), T. Munshi (CEPT University, Ahmedabad), S. Purkayastha (ISI, Kolkata, India), K. Rajendran (IISc., Bangalore), S. Sikdar (George Mason University, Washington D. C.), L. P. Singh (Hyderabad), J. S. Steyer (Paris) and others for reviewing the manuscripts and their constructive suggestions. I am grateful to the authors for their cooperation. I thank D. P. Sangupta and C. Chakraborty and my other colleagues for their help and encouragements.

P. GHOSH

List of Contributors

Abdul Matin
Department of Geology
Calcutta University
Kolkata, India

Alok Porwal
Centre for Exploration Targetting
The University of Western Australia
Crawley, WA, Australia

Alok Porwal
State Department of Mines & Geology
Rajasthan, Udaipur, India

Asit Saha
Geodata and Database Division
Geological Survey of India
27, Jawaharlal Nehru Road
Kolkata-700016, India

Atin Kumar Mitra
Department of Geological Sciences
Jadavpur University, Kolkata-700032, India

Auditeya Bhattacharyya
Geodata and Database Division
Geological Survey of India
27, Jawaharlal Nehru Road
Kolkata-700016, India

Barendra Purkait
Publication Division
Geological Survey of India
29, Jawaharlal Nehru Road
Kolkata-700016, India

Basab Mukhopadhyay
Geodata and Database Division
Geological Survey of India
27, Jawaharlal Nehru Road
Kolkata – 700016, India

Bikash C. Poddar
Center for Study of Man and
Environment (CSME), CK-11
Sector-2, Salt Lake, Kolkata-700091,
India

Dhruba Mukhopadhyay
Department of Geology
University of Calcutta
Kolkata-700019, India

Dhurjati P. Sengupta
Geological Studies Unit
Indian Statistical Institute
203 Barrackpore Trunk Road
Kolkata 700108, India

E.J.M. Carranza
International Institute for Geo-information
Science and Earth Observation (ITC)
Enschede, The Netherlands

M. Hale
Faculty of Geosciences
Utrecht University, Utrecht
The Netherlands

M. Hale
International Institute for Geo-information
Science and Earth Observation (ITC)
Enschede, The Netherlands

Malay Mukul
CSIR Centre for Mathematical
Modelling and Computer Simulation
Bangalore, India

Mallika Mullick
Center for Study of Man and
Environment (CSME), CK-11
Sector-2, Salt Lake, Kolkata-700091
India

Manhal Mahmoud
Department of Geological Sciences
Jadavpur University, Kolkata-700032
India

Nibir Mandal
Indian Institute of Science Education
and Research, HC-7, Salt Lake
Kolkata-700106, India

Parthasarathi Ghosh
Geological Studies Unit, Indian Statistical
Institute, 203 Barrackpore Trunk Road
Kolkata-700108, India

Ritabrata Dutta
Geological Studies Unit, Indian Statistical
Institute, 203 Barrackpore Trunk Road
Kolkata-700108, India

Shamik Sarkar
Indian Institute of Science Education
and Research, HC-7, Salt Lake
Kolkata-700106, India

Sridevi Jade
CSIR Centre for Mathematical Modelling
and Computer Simulation, Bangalore
India

❑❑❑

CONTENTS

CHAPTER 1

Active Deformation in the Darjiling-Sikkim Himalayas based on 2000-2004 Geodetic Global Positioning System Measurements

Malay Mukul, Sridevi Jade, Abdul Matin

Abstract

High precision Geodetic Global Positioning System measurements have become important for studying present-day tectonics in active orogenic belts such as the Himalayas. Geodetic Global Positioning System measurements in the Darjiling-Sikkim Himalayas (DSH) during 2000-2004 indicate that the frontal part of the Darjiling Himalaya is locked to the Indian plate and statistically negligible shortening is accumulating there at present. While this observation is in conformity with observations made in Nepal to the west of the DSH, it differs in the fact that about ~ 12 mm/yr of convergence is accommodated in the DSH and the western Bhutan Himalayas as opposed to about ~ 20 mm/yr in Nepal Himalaya. About 16 mm of convergence is accommodated in the Arunachal Himalaya, to the east of DSH, of which about 6 mm is accommodated in the Lesser Arunachal Himalaya. These observations indicate that there is a fair amount of heterogeneity in the way convergence is accommodated along the length of the Himalayan arc and the Himalayan seismic hazard estimations need to account for this. Seismicity patterns in the DSH indicate that active deformation in the area is driven by strike-slip as well as thrust tectonics. Strain accumulation in the DSH during

2000-2004 appears to be dominated by strike-slip tectonics in the region and about ~ 4-5 mm/yr of sinistral strike-slip is observed along a NNE-SSW trending near-vertical fault identified in the area (named as the Gish transverse fault).

1.1 Introduction

The Global Positioning System (GPS) is a navigation and precise-positioning tool developed by the United States Department of Defense in 1973 primarily for military use. Subsequently, it was recognized that the precise-positioning provided by GPS could be used in high-precision geodesy and for continental deformation studies to measure tectonic motions during and in between earthquakes (e. g. Bilham and Gaur, 2000 and references therein) and provide data on the present day deformation in orogenic belts that have been active over geological time scales (e.g. Mukul, 2005 and references therein). This paper presents an introductory review on the use of high precision GPS as a tool for understanding present-day deformation in the Indian plate and then presents the results of a new GPS based study from the Darjiling-Sikkim Himalaya that points to a fair amount of heterogeneity in the way convergence is accommodated along the length of the Himalayan arc.

1.1.1 The Global Positioning System

Three distinct segments make up the Global Positioning System (Hoffmann-Wellenhof *et al.,* 1997). The first segment of the system consists of 24 satellites, orbiting 20,000 km above the Earth in 12-hour circular orbits. In order for the satellites to be detected from anywhere on the earth's surface, they are divided into six groups of four that follow different assigned paths thereby creating six orbital planes that completely surround the earth. These satellites send radio signals to earth that can be detected using ground-based GPS receivers. The second segment of the GPS system is the ground station comprising of receiver(s) and antennae together with communication tools to transmit data to a data centre. The omni-directional antenna at each site picks up the satellite signals and transmits them to the site receiver as electric currents. The receiver separates the signal into different channels designated for a particular satellite and frequency at a particular time. After the signals have been isolated, the receiver can decode them and split them into individual frequencies and produce a general position (latitude, longitude, and height) for the antenna with this information. The third segment of the system monitors and controls the Global

Positioning System and the International Global Navigation Satellite Systems Service (IGS) stations. It uses automated computer systems to retrieve and analyze data from the receivers at these stations. Once processed, the data is made available for use on the internet. The IGS stations are treated as reference stations whose positions are, for all practical purposes, known.

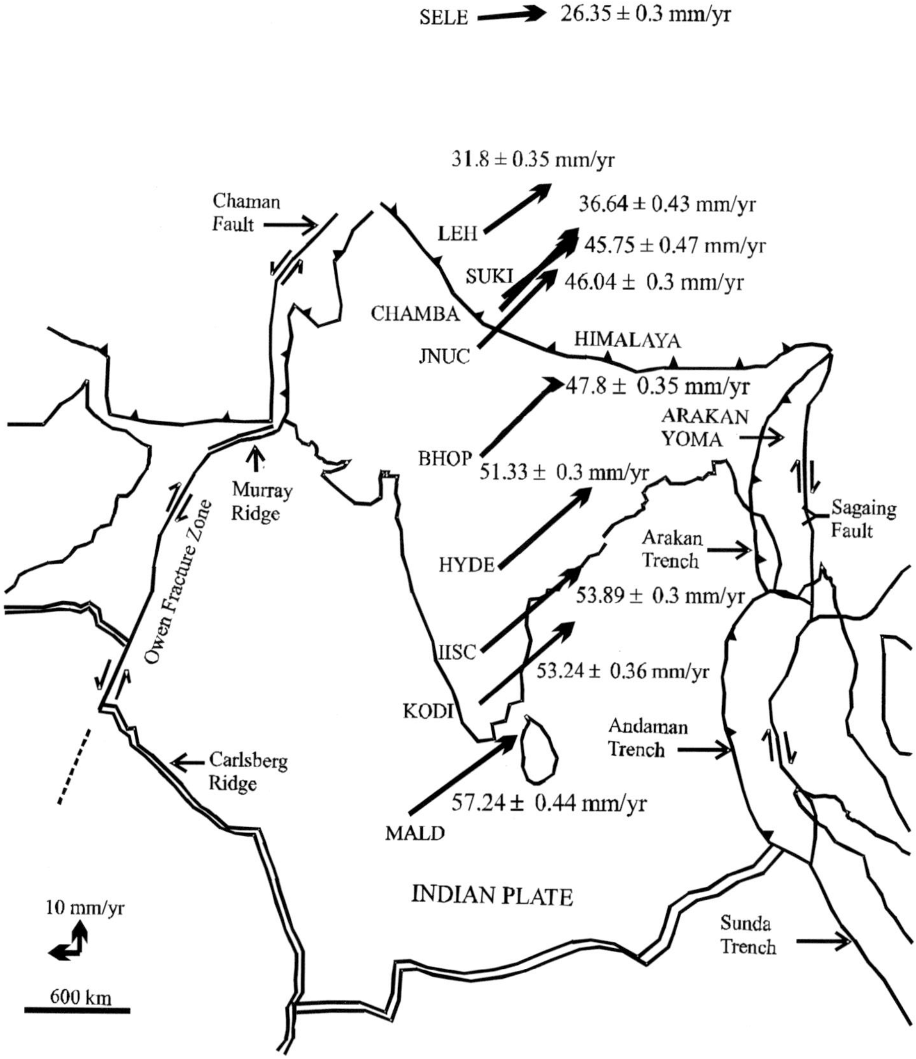

Fig. 1.1 : The Indian plate and plate boundaries along with variation in ITRF 2000 GPS velocities from its southern divergent boundary to the northern convergent boundary in the Himalayas. Stations are located at the base of the arrows.

1.1.2 Active Deformation in the Indian Plate using Global Positioning System

The present-day Indian plate is defined by the region between the Carlsberg Ridge in the Indian Ocean (the southern divergent plate boundary) and the Himalaya (northern convergent plate boundary) (Fig. 1.1). The eastern and western boundaries of the present-day Indian plate are more complex. The western boundary is a composite boundary. It is defined by the Owen fracture zone connecting the Carlsberg ridge with Murray Ridge and the sinistral, transform, Chaman fault connecting the Murray ridge with the Himalaya (Fig. 1.1). The eastern boundary is defined by the dextral Sagaing fault and the Sunda-Arakan trench (Fig. 1.1) that extends into the Indo-Burman region as the Indo-Burman Fold Thrust Belt (IBFTB) in the Arakan Yomas. The Chaman fault exhibits aseismic sinistral slip of ~ 26 mm/yr (Apel *et al.,* 2006) whereas the Sagaing fault exhibits dextral slip of ~ 20 mm/yr (Vigny *et al.,* 2003).

Given that the Indian plate motion occurs on a spherical earth, the motion is characterized by an Euler pole of rotation and an angular velocity corresponding to rotation about the pole. Several Euler poles of rotation have been computed by different authors using different sets of stations on the Indian plate (Table 1.1) in the ITRF 2000 reference frame. The composite Indian ITRF 2000 Euler pole computed using all the estimated Euler poles (Table 1.1) is determined as 51.33 ± 1.79 N; 12.7 ± 1.94 W. The composite angular velocity of the Indian plate in ITRF 2000 reference frame is 0.479 ± 0.006 °/ Myr.

Table 1.1: Euler Poles of Rotation in ITRF 2000 reference frame for Indian Plate Motion.

Euler Pole	Latitude (° N)	Longitude (° E)	Angular Velocity (°/Myr)
Jade *et al.,* 2007	51.7 ± 0.5	-15.1 ± 1.5	0.469 ± 0.01
Bettinelli *et al.,* 2006	51.4 ± 1.6	-10.9 ± 5.6	0.483 ± 0.01
Socquet *et al.,* 2006	50.9 ± 5.1	-12.1 ± 0.6	0.486 ± 0.01
Composite Pole	51.3 ± 1.7	-12.7 ± 1.9	0.479 ± 0.01

The velocity within the Indian plate is expected to vary from south to north, i. e. from the divergent to the convergent margin. Typically, plate velocities are expected to be maximum at the divergent margins and minimum at the convergent margins. Therefore, the nature of variation in the plate velocity within a given plate from one boundary to the other determines the regions of strain accumulation in the plate. Plate velocities measured in the Indian plate (in the ITRF 2000 reference frame) from south to north (approximately along the 77-78E longitudes) vary (Jade *et al.*, 2007) from ~ 57 mm/year in the Maldives (04.19 N) to ~ 46 mm/yr in Delhi (28.54 N) gradually. It remains ~ 46 mm/yr at Chamba (30.404 N), (Fig. 1.1) but is ~ 37 mm/yr at Suki (30.997 N) (Jade, 2004) indicating abrupt slowing down of the Indian plate between these two stations. The velocity drops to ~ 32 mm/yr at Leh (34.128 N) (Jade *et al.*, 2004) indicating that the Indian plate slows down by ~ 25 mm/yr from near its southern divergent boundary to its northern convergent boundary; the maximum deceleration of the Indian plate occurs in the higher Himalayan region north of 30 N latitude (Fig. 1.1). This region is also likely to accommodate maximum strain because of this deceleration. All velocity vectors in the Indian plate are directed to the north-east in the ITRF 2000 reference frame whereas velocity vectors in the Eurasian plate near the Himalayan boundary point eastwards (SELE 43.179 N). This indicates the existence of a region in or north of the Himalaya where the velocity vectors switch from a dominantly north easterly to an easterly trend in the ITRF 2000 frame.

The variation in velocity from the divergent to the convergent boundary in the Indian plate gives rise to deformation in the Indian plate that can be quantified by looking at north-south baseline changes within the Indian plate (Fig. 1.1; Table 1.2). The Maldives-Bangalore and Bangalore-Delhi baselines exhibit 2.14 ± 0.23 and 2.2 ± 0.1mm/yr of shortening respectively thereby indicating a total of ~ 4 mm/yr shortening from the southernmost point, Maldives to Delhi in northern India. This observation is also supported by 2.4 ± 0.2 mm/yr shortening along the Bangalore-Chamba baseline. Also, the Delhi-Leh and the Chamba-Suki baselines exhibit 12.8 ± 0.1 and 10.51 ± 0.41 mm/yr of shortening respectively indicating that ~ 12 mm/yr of shortening is taken up in this part of the Himalaya in contrast with ~ 3 mm/yr shortening in peninsular India. An additional 16.9 ± 0.05 mm/yr shortening is taken up north of Leh and the Himalaya along the Leh-Almaty baseline.

1.1.3 Active Tectonics in the Himalaya using Global Positioning System

GPS measurements along the length of the Himalayan arc (Fig. 1.2) have been used to compute variations in convergence rates along the length of the mountain belt. For example, about 15 mm/yr convergence is being accommodated in the Ladakh Himalayas (Banerjee and Burgmann, 2002, Jade *et al.*, 2004). Farther east, in the Garhwal Himalayas 13.8 ± 2.2 mm/yr convergence is being accommodated between Delhi (28.54 N, 77.17 E) and Suki (30.99 N, 78.68 E) (Jade, 2004). In the Kumaon Himalayas, the convergence between Delhi (28.54 N, 77.17 E) and Munsiari (30.06 N, 80.19 E) has been measured to be 12.2 ± 2.2 mm/yr (Jade, 2004). Farther to the east in central Nepal, GPS measurements have yielded ~20 mm/yr convergence (Bilham *et al.*, 1997; Jouanne *et al.*, 1999; Larson *et al.*, 1999). Approximately, 16 ± 0.5 mm/yr is being accommodated in eastern Bhutan between Guwahati (26.11N, 91.47 E) and Lhasa (29.65 N, 91.10E) (Jade *et al.*, 2007). Again, 15 ± 0.1 mm/yr is being accommodated in western Arunachal Himalaya between Tezpur (26.62 N, 92.78 E) and Lhasa (29.65 N, 91.10 E) (Jade *et al.*, 2007). These results seem to indicate that there is variation in the shortening that is accommodated along the Himalayan arc; the maximum shortening is accommodated in the central part of the arc in Nepal.

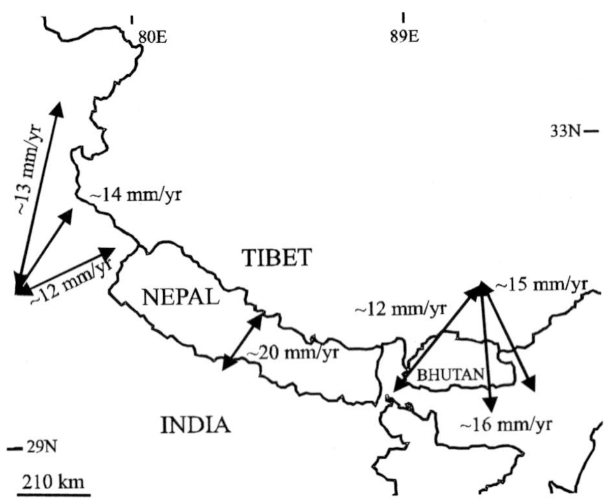

Fig. 1.2 : Shortening estimates in the Himalayan arc from GPS measurements indicate that the maximum shortening is accommodated near the central part of the arc.

1.1.4 Regional Geology of the Darjiling-Sikkim Himalayas

The Darjiling-Sikkim Himalayas (DSH) (Acharyya, 1994, Ganser, 1983, Mukul, 2000, Nakata, 1989, Schwan, 1980) lies between the Nepal - India border to the west and the Bhutan Himalaya to the east (Gansser, 1983) (Fig. 1.3). Relationship between the major faults in

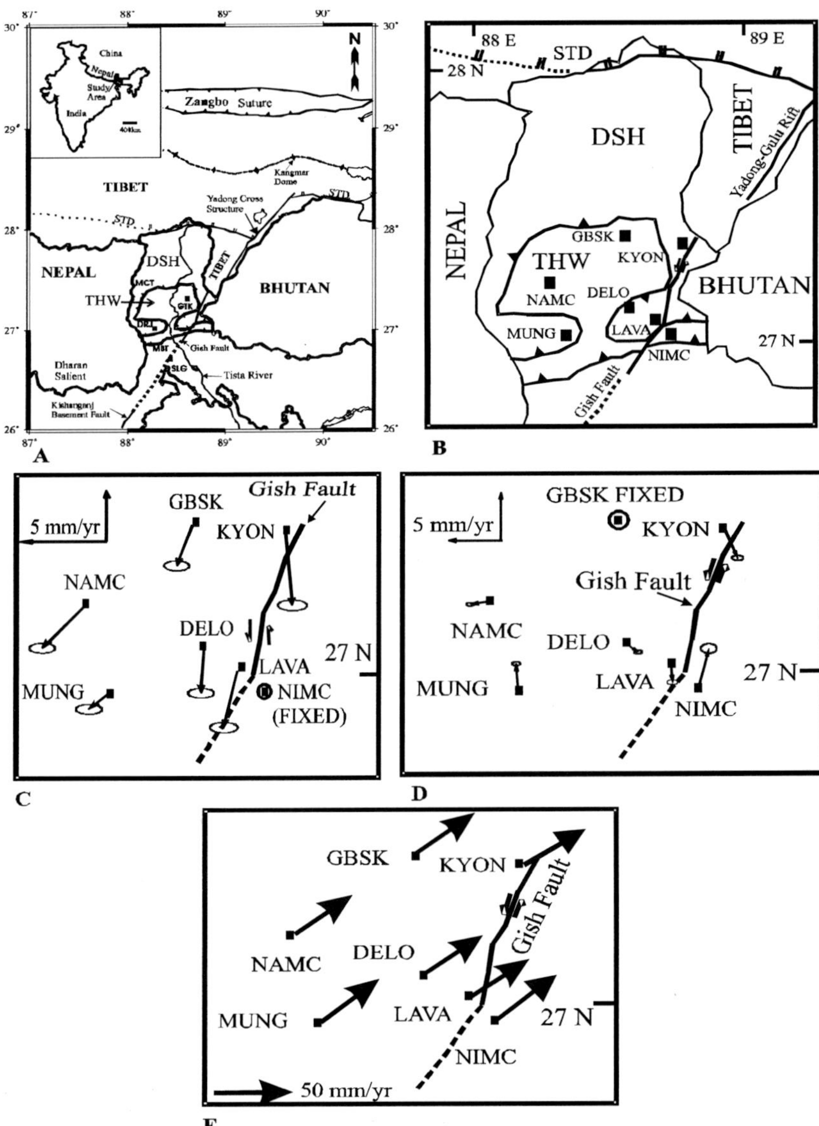

Fig. 1.3 : GPS stations and velocity vectors in the Darjiling-Sikkim Himalayas (DSH). A map of the GPS stations in relation with regional structure in the area is shown in (A). The Tista half-window (THW) and the Gish Transverse Fault (GTF) trace along with the GPS stations are shown in (B). Southward motion of the west block of the GTF relative to the east block is evident from velocity vectors plotted relative to NIMC (C). Northwards motion of the east block relative to the west block of the GTF is seen from velocity vectors plotted relative to the permanent station GBSK (D). The locked part of the GTF has been dashed. Slip of around 4-5 mm/yr is observed along the GTF. Velocity vectors in ITRF 2000 reference frame are shown in (E). The uncertainties associated with the velocity vectors in (E) are negligible compared to their lengths (Table 1.5).

the Himalayas such as the Main Central Thrust (MCT) and the Main Boundary Thrust (MBT) persist in the Darjiling-Sikkim Himalayan fold-and-thrust belt (FTB). However, the lithounits associated with the Himalayan thrusts in the Darjiling-Sikkim FTB are vertically stacked, with the higher grade thrust sheets transported farther southward than elsewhere in the Himalaya. This results in the MCT and the MBT surface traces being exposed within a few kilometres of each other in the frontal part of the mountain belt (Fig. 1.3a & b). The Tista half window (Fig. 1.3a & b) reveals that the MCT is eroded and the MCT sheet rocks are underlain by the Proterozoic Buxa group, Daling-Reyang schists (Table 1.3) (e. g. Acharyya, 1980) carried by the North Kalijhora Thrust (NKT) (Mukul, 2000). The MBT carries the Gondwana miogeoclinal flysch and molasses (Acharyya, 1976) and underlies the Proterozoic NKT sheet as evident from an additional window in the Tista half-window in the Rangit valley (Raina, 1976) along the crest of the Rangit anticline (Schwan, 1980, Ray, 2000). In the frontal portion of the Tista half-window, the MBT carries Permian Gondwana rocks over Tertiary Siwalik section (Table 1.2). The MBT is folded by the South Kalijhora thrust (SKT) (Mukul, 2000, Basak and Mukul, 2000). The structure in the footwall of the SKT is described as an imbricate thrust zone that repeats the Middle Siwalik Geabdat sandstone section (Table 1.2) several times (Acharyya, 1994, Mukul, 2000, Basak and Mukul, 2000); successive imbricates extend for over 20 km in front of the MBT and some of the frontal imbricates are probably blind. The mountain front to the south of the Tista half-window was in place by ~ 40 ka (Mukul *et al.,* 2007).

The mountain front in the DSH is sinuous in map-view and defines salient-recess geometry (Fig. 1.4). The River Tista exits the mountain front in the eastern part of the Dharan salient. East of the Tista half window, the mountain front recedes by ~10 km and defines the Gorubathan recess. The transition zone between the eastern part of the Dharan salient and the Gorubathan recess is defined by a transverse fault. The transverse fault has been mapped as an unnamed fault (Acharyya, 1971) along the Lethi River near the DSH mountain front. There are no Siwalik exposures in the Gorubathan recess in the DSH and the geomorphic contrast between the Siwalik Hills to the west and alluvial plains to the east of the transverse fault is prominent (Fig. 1.4 inset). The linear nature of the contact and the geomorphic contrast between the rugged and forested Siwalik Hills to the west and flat, cultivated, plains to the east (Fig. 1.4) suggests the presence of a NNE-

SSW trending, high-angle fault. The Gish-Lethi River follows the trend of the transverse fault zone and, therefore, the fault is named as the Gish Transverse Fault (GTF) (Fig. 1.4).

Table 1.2 : Stratigraphy exposed in the frontal Darjiling-Sikkim Himalayas after Mukul (2000) and Nakata (1989) and the references therein.

Age	Stratigraphic unit	West of Gish Transverse Fault	East of Gish Transverse Fault
Cenozoic			
Mio-Pliocene	Upper Siwalik	Murti boulder bed Crude bedded immature conglomerate Parbu grit, Pebbly sandstone and coarse to medium sandstone	Not exposed
	Middle Siwalik	Geabdat medium to coarse - grained sandstone and shale, local pebbly beds, minor marl	Not exposed
	---------------	South Kalijhora Thrust (SKT)	----------------------
	Lower Siwalik	Chunabati Formation Fine-to medium-grained sandstone, siltstone, claystone, marl, basal conglomerate	Not exposed
------------	---------------	Main Boundary Thrust (MBT)	----------------------
Paleozoic			
Upper Permian	Gondwana Group	Damuda Subgroup Sandstone, carbonaceous shale and coal	Damuda Subgroup Sandstone, carbonaceous shale and coal
Lower Permian	Gondwana Group	Rangit pebble-shale (Talchir?) Diamictite, rythmite, quartzite marl	Not exposed
------------	---------------	North Kalijhora Thrust (NKT)	Gorubathan-Jiti Thrust
Pre-Cambrian	Daling Group	Buxa Formation Dolomites, fine-grained quartzites and pyritiferous shales	Not exposed
		Reyang Formation Variegated quartzites, shales and slates	Not exposed
		Daling Formation	Gorubathan Formation

Contd...

	Chlorite-sericite, greenish phyllites, Quartzite and slates, interbanded phyllites, Quartzite and metabasics slates, interbanded metabasics	Chlorite-sericite, greenish phyllites, Quartzite and slates, interbanded phyllites, Quartzite and metabasics slates, interbanded metabasics
----------	Main Central Thrust I (MCT I)Paro-Lingste biotite schists and gneisses	Paro-Lingste biotite schists andgneisses
----------	Main Central Thrust II (MCT II) High grade gneisses	High grade gneisses

The GTF has been mapped from the mountain front to the Indo-Tibet border over an aerial distance of ~50 km as a ~1.5 km wide, NNE-SSW trending and steeply dipping zone that shows extensive elastico-frictional fault zone deformation features (Mukul and Matin, 2005). Regional sense of displacement on the GTF was difficult to determine due to the lack of suitable offset markers. However, the traces of the east-west regional thrusts and associated lithocontacts were seen to be displaced to the north (Acharyya, 1971) close to the mountain front in the east block of the GTF relative to the west block (Fig. 1.4) indicating sinistral motion. Therefore, the GTF is a sinistral, near-vertical to steeply west-dipping strike-slip fault (Mukul and Matin, 2005) that extends from the mountain front northward to the Indo-Tibet-Bhutan border, thereby joining the Yadong cross-structure in Tibet (Fig. 1.4).

1.1.5 Seismotectonics in the Darjiling-Sikkim Himalaya

No great earthquake has been reported in recent or historic past in the DSH. The seismicity in the DSH is dominated by frequent moderate earthquakes (De, 1996, 2000, Kayal, 2001, De and Kayal, 2003, 2004, Nath *et al.,* 2005). The earthquake epicenters are more-or-less confined to the region between latitudes 27° N and 27.5° N (Fig. 1.5). The Tista half-window records majority of the earthquakes in the area and very few epicenters plot north of the MCT trace in the Tista half-window. N-S depth sections across the Tista half-window (De, 2000, Kayal, 2001, De and Kayal, 2003, Nath *et al.,* 2005) reveal that many hypocenters extend to depths of 45 km prompting interpretations of an unusually deep MBT (Fig. 1.6). We contend that the deep earthquakes in the Tista half-window may be genetically related to the NNE-SSW trending and steep, west-dipping GTF as deep

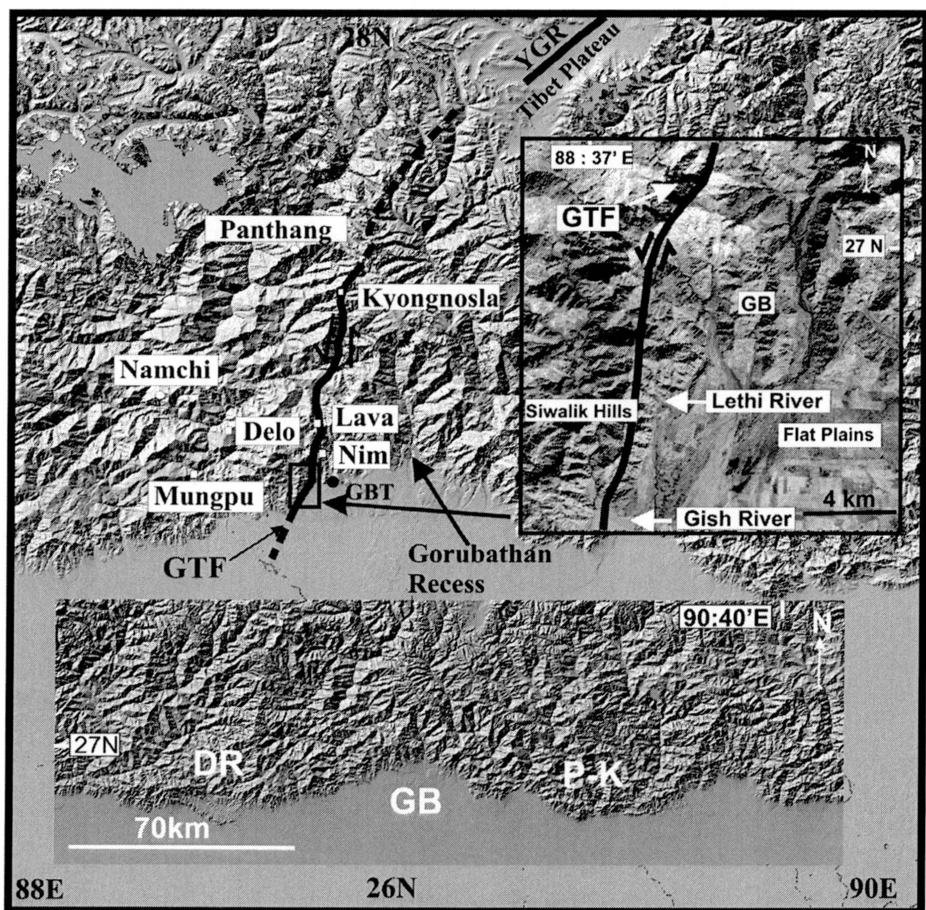

Fig. 1.4 : SRTM (Shuttle Radar Topography Mission) imagery of the Darjiling Sikkim Himalayas showing the Gorubathan recess, the surface trace of the GTF, the Tibet plateau and the Yadong-Gulu Rift (YGR). Location of the permanent station at Panthang (GBSK) and campaign mode GPS stations in the Darjiling-Sikkim Himalayas at Labha (LAVA), Nim (NIMC), Kyongnosla (KYON), Delo Hill (DELO), Namchi (NAMC) and Mungpu (MUNG) are shown. The GTF has a prominent geomorphic expression in the frontal part of the area (right inset) and separates low Siwalik hills on the west from flat cultivated plains to the east. The GTF has developed in the transition zone between the Dharan salient (DR) and the Gorubathan recess (GB). The Gorubathan recess is located between Dharan salient and the Phuntsoling-Kalikhola (P-K) salient in Bhutan (bottom).

earthquake hypocenters seem to be confined to west of the surface trace of the GTF. This contention is supported by the fact that an unusually deep MBT is not consistent with the data from east of the GTF (De, 1996) where MBT is interpreted to be a listric fault that joins the basal Himalayan decollement (De, 1996). Also, the seismicity observed in the Tista half-window west of the surface trace appears to be related to two mechanisms (Fig. 1.4; De and Kayal, 2004, Ni and Barazangi, 1984); active thrust faulting related to the E-W thrust faults and sinistral strike-slip faulting (Fig. 1.3) that can be related to the NNE-SSW trending sinistral GTF. Fault plane solutions of strike-slip faulting (Fig. 1.5) support such a contention as one of the focal planes trend NNE-SSW and show sinistral strike-slip motion. We, therefore, interpret the GTF to be a seismogenic, sinistral, near-vertical to steep-dipping strike-slip fault in the transition zone between the Dharan salient and the Gorubathan recess that segments the Himalayan arc in the Darjiling-Sikkim region (Fig. 1.4).

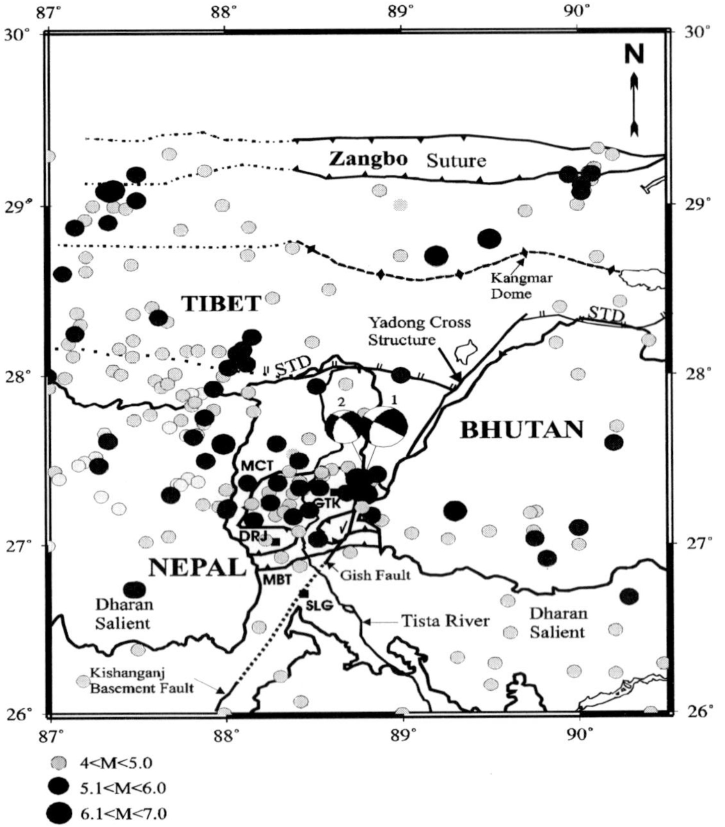

Fig. 1.5 : Map showing the main faults, distribution of earthquake foci between 1800 and 2006 (Sources: ISC, NOAA, IMD, NEIC) in the Darjiling-Sikkim and surrounding area. Focal Mechanisms of two moderate (Table 1.8) strike-slip earthquakes near Kyongnosla are shown and are interpreted to be related to the GTF. The foci occur within the Tista half-window indicating that the seismicity occurs below the structural level of the Main Central Thrust (MCT).

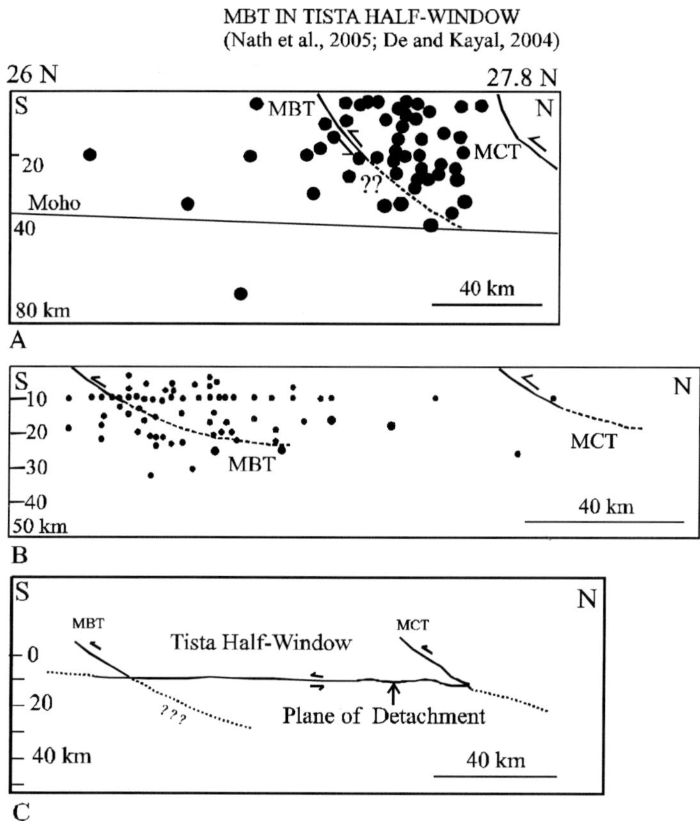

Fig. 1.6 : Seismicity in the Tista half-window showing deep hypocenters below the plane of detachment (after De and Kayal, 2004 and Nath *et al.*, 2005) in (A) and (B). The Main Boundary thrust (MBT) has been interpreted to be the source of these earthquakes which implies that it extends below the basal decollement (Main Himalayan Thrust) in the region (C). We contend that these hypocenters are related to the GTF and the MBT soles with the decollement like elsewhere in the Himalayas.

1.2 Geodetic Global Positioning System Measurements in the Darjiling-Sikkim Himalaya

GPS measurements were first initiated in the DSH (Fig. 1.3) in 1997 at Delo Hill (Table 1.3) near Kalimpong town. Subsequently, 4 sites were added in 2000 at Labha, Mungpu, Kyongnosla, and Namchi. The first order network was completed in 2002 with the establishment of the Nim station. In addition to these campaign sites, a permanent station was established at Panthang (Gangtok) in November 2003. All the stations were equipped with Trimble 4000, 5700, Leica SR520, and RS500 GPS receivers with choke ring and Zephyr geodetic antennas. The permanent station antenna was mounted on a concrete pillar that is grouted to bedrock. Campaign sites were marked by 2 to 3mm diameter holes drilled on bedrock exposures to allow subsequent re-occupations on bipods. The results presented here are based on continuous data from IGS stations and GPS campaign measurements carried out in the DSH during 2000-2004. In addition, continuous data from the permanent station (Panthang) during 2003-2004 has also been used.

Table 1.3 : GPS Stations used in the analysis.

Site	Code	Lat (° N)	Long (° E)	GPS Measurements	Information about the site
Kyongnosla (Campaign)	KYON	27.37	88.71	308-310 (2000) 84, 356-358 (2002) 305-308 (2003) 82-85 (2005)	Located in high grade gneiss in west block of GTF
Nim (Campaign)	NIMC	27.00	88.69	362-364 (2002) 306-308 (2003) 356-358 (2004)	Located in Daling schists in east block of GTF
Labha (Campaign)	LAVA	27.07	88.66	307-310 (2000) 361-363 (2002) 306-308 (2003) 359-362 (2004)	Located in Paro biotite schists in west block of GTF
Delo hill (Campaign)	DELO	27.09	88.50	343-351 (1997) 281-283 (1999) 308-315 (2000) 89-91, 355-361(2002) 310-312 (2003) 353-356 (2004)	Located in Paro biotite schists in west block of GTF
Mungpu (Campaign)	MUNG	26.98	88.40	312-315 (2000) 82-83 (2002) 310-312 (2003) 353-357 (2004) 80-82 (2005)	Located in Daling schists
Namchi (Campaign)	NAMC	27.16	88.32	312-315 (2000) 78-80 (2002) 310-312 (2003) 83-86 (2005)	Located in Daling schists
Panthang, Gangtok (Permanent)	GBSK	27.37	88.57	305, 2003-362 (2004)	Established in November 2003 in Daling schists.

Contd...

Kunming	KUNM	25.03	102.79	343-351 (1997) 281-283 (1999) 307-315 (2000) 78-91, 356-364 (2002) 305, (2003) - 362, (2004)	IGS Station, KUNM
Lhasa	LHAS	29.66	91.10	343-351 (1997) 281-283 (1999) 307-315 (2000) 78-91, 356-364 (2002) 305, (2003) - 362, (2004)	IGS Station, LHAS
Hyderabad	HYDE	17.41	78.55	305, (2003) - 362, (2004)	IGS Station, HYDE
Bangalore	IISC	13.02	77.57	343-351 (1997) 281-283 (1999) 307-315 (2000) 78-91, 356-364 (2002) 305, (2003) - 362, (2004)	IGS Station, IISC
Almaty	SELE	43.18	77.02	343-351 (1997) 281-283 (1999) 307-315 (2000) 78-91, 356-364 (2002) 305, (2003) - 362, (2004)	IGS Station , SELE
Bishkek	POL2	42.68	74.69	343-351 (1997) 281-283 (1999) 307-315 (2000) 78-91, 356-364 (2002) 305, (2003) - 362, (2004)	IGS Station, POL2
Maldives	MALD	4.19	73.53	305, (2003) - 362, (2004)	IGS Station, MALD
Kitab	KIT3	39.14	66.89	343-351 (1997) 281-283 (1999) 307-315 (2000) 78-91, 356-364 (2002) 305, (2003) - 362, (2004)	IGS Station, KIT3
Bahrain	BAHR	26.21	50.61	343-351 (1997) 281-283 (1999) 307-315 (2000) 78-91, 356-364 (2002) 305, (2003) - 362, (2004)	IGS Station, BAHR

1.2.1 Data Processing and Analysis

GPS data collected at all the above sites were first converted into rinex observation files and initial quality check was performed using TEQC (Translations, Editing and Quality Checking Software). The quality check plots of all the GPS data were carefully examined and the data containing high cycle clips, multipath and < 12hr observations were removed from the analysis. The GPS campaign style data (1997-

2004) were processed along with the permanent station data (Fig. 1.3) and the IGS stations (Table 1.3) using the GAMIT/GLOBK processing software developed by Massachusetts Institute of Technology (MIT), USA (King and Bock, 2000). Minimum 2 to 3 days (24 hr files) of data per epoch for campaign sites and continuously available daily data of all permanent sites (Table 1.3) with sampling interval of 30 sec and satellite elevation cut off angle of 15° were used in the data analysis. Daily loose GAMIT solutions accounted for error contributions due to signal path delay caused by atmosphere, orbital accuracy, antenna phase centre variations, signal multi-path and scattering by the receiver environment, satellite and receiver clock errors. Ambiguity free and ambiguity fixed solutions were performed with ionosphere free linear combination to account for carrier phase ambiguities and signal delay due to ionosphere. Zenith tropospheric delay for each station was estimated by incorporating a piecewise-linear model with stochastic constraints and then corrected for the signal delay due to troposphere.

The daily coordinates and velocities of all permanent and campaign sites were estimated in the ITRF 2000 reference frame (Altamimi *et al.*, 2002) by constraining IGS reference station positions and velocities in the region to reported values in that frame with standard errors provided by IGS. The horizontal error ellipses were computed by GAMIT/ GLOBK software from the uncertainties in the North and East coordinates / velocities using the correlation between the baseline component estimates.

1.2.2 Active Tectonics in the Darjiling-Sikkim Himalaya from GPS

GPS measurements allow the estimation of very short-term, active deformation starting from the day a station is set-up and measured for the first time. The Bangalore (IISc: 13.02° N; 77.57° E) and Lhasa (LHAS: 29.66° N; 91.10° E) are two permanent IGS stations that are relevant to the study of the deformation in the regions because the IISc-Lhasa baseline is a good proxy for deformation in the Darjiling-Sikkim and Western Bhutan Himalayas. The IISc velocity (51.97 ± 0.63 mm/yr; N 51.75 ± 0.63 E) is representative of the Indian plate whereas the Lhasa velocity (46.2 ± 0.59 mm/yr; N 74.71 ± 0.23 E) is indicative of predominantly easterly motion characteristic of the Tibetan plate. The ITRF 2000 model for IISc and Lhasa IGS stations predicts a shortening of ~10 mm/year of the IISc-LHAS baseline for the period 1997-2004 (Table 1.4). The computed IISc-LHAS baseline shortening for the period 2000-2004 is 11.93 ± 0.34 mm/yr (Table 1.5) indicating that less shortening (~ 12 mm/yr) is accommodated in the Darjiling-Sikkim-western Bhutan Himalaya compared to its west (~ 20 mm/yr; Bilham *et al.,* 2001) in central Nepal and east (~ 15 mm/yr; Jade *et al.,* 2007) in eastern Bhutan.

Table 1.4 : The International Terrestrial Reference Frame (ITRF) 2000 based convergence computation between IISc (Bangalore) and LHAS (Lhasa). ITRF 2000 Co-ordinates and Velocities on January 1, 1997 (http://igs.ifag.de/root_ftp/ITRF2000/ITRF2000_GPS.SSC). IISc-LHAS ITRF 2000 Baseline length on January 1, 1997 was 2299529.25542 ± 0.0027m. ITRF 2000 velocities of the two stations indicate a convergence rate of 10.18 mm/yr between these two IGS stations.

Station	X (m)	Y (m)	Z (m)	VX (m/yr)	VY (m/yr)	VZ (m/yr)
IISc	1337936.813 ± 0.002	6070317.106 ± 0.003	1427876.516 ± 0.002	-0.041 ± 0.0009	0.003 ± 0.0007	0.0334 ± 0.0005
LHAS	-106937.669 ± 0.001	5549269.591 ± 0.003	3139215.762 ± 0.002	-0.0462 ± 0.0005	0.0076 ± 0.0011	0.0121 ± 0.0007

The velocities measured in the frontal part of the Darjiling-Himalaya at Mungpu and Nim, both west and east of the GTF (Table 1.5) indicate that the Indian plate slows down to ~ 49 mm/yr and ~ 51 mm/yr respectively from ~ 52 mm/yr at Bangalore. This indicates very small amounts of shortening (Table 1.6; IISc-Mungpu (~ 3mm/yr) and IISc-Nim (~ 2 mm/yr)) in eastern peninsular India. The campaign station at Mungpu and the permanent station at Panthang are located on the south and north ends of the Tista half-window to measure the N-S shortening across the half-window where maximum seismicity is observed in the DSH. The velocity at Panthang is only ~ 1.5 mm/yr lower than at Mungpu (Table 1.5) which indicates that not much strain is accumulating in the Tista half-window either; the Mungpu-Panthang baseline shows shortening of about 2 mm/yr. Both these stations are located in the west block of the GTF. They are also located away from the GTF fault zone (Fig. 1.3). Therefore, they are likely to represent shortening that is being accommodated along active east-west faults in the Tista half-window. Kyongnosla, however, is located very close to the GTF fault zone in the west block of the fault and the Mungpu-Kyongnosla baseline shows a shortening of ~ 4 mm/yr. Therefore, a maximum of ~ 4mm/yr shortening is accommodated in the Tista half-window west of the surface trace of the GTF (Table 1.6). However, quantification of present-day slip on the seismogenic

GTF was the main objective of the GPS measurements in Darjiling-Sikkim Himalaya and we discuss that in detail now.

Table 1.5 : Coordinates and ITRF 2000 Velocities of GPS Stations during 2000-2004 relevant to the tectonics of the study region.

Station (* Permanent,^ IGSStation)	Latitude (°N)	Longitude (°E)	Velocity(E) (mm/yr)	Velocity (N) (mm/yr)	Composite Velocity (mm/yr)
IISC^	13.021	77.570	40.82 ± 0.71	32.16 ± 0.46	51.97 ± 0.63 N51.75 ± 0.63E
LHAS^	29.657	91.104	44.57 ± 0.59	12.16 ± 0.57	46.20 ± 0.59 N74.71 ± 0.23E
GBSK* (2003-2004)	27.365	88.569	38.10 ± 0.82	28.39 ± 0.60	47.51 ± 0.75 N53.29 ± 0.83E
KYON	27.367	88.714	39.42 ± 1.09	25.19 ± 0.69	46.78 ± 0.99 N57.4 ± 1.01E
NAMC	27.157	88.322	36.51 ± 0.78	28.11 ± 0.58	46.08 ± 0.71 N52.39 ± 0.82E
DELO	27.089	88.503	38.85 ± 0.7	27.81 ± 0.56	47.78 ± 0.66 N54.38 ± 0.73E
LAVA	27.070	88.664	38.25 ± 0.98	26.89 ± 0.66	46.76 ± 0.89 N54.87 ± 0.96E
NIMC (2002-2004)	26.998	88.685	39.03 ± 1.54	32.41 ± 1.09	50.73 ± 1.37 N50.27 ± 1.46E
MUNG	26.978	88.400	37.95 ± 0.82	30.90 ± 0.59	48.94 ± 0.74 N50.83 ± 0.81E

1.2.3 Quantification of Present-day Slip Rate on the Gish Transverse Fault (GTF)

Five campaign mode stations were set up to determine the slip on the GTF. The campaign mode stations at Labha (LAVA: 27.07° N, 88.66° E) and Kyongnosla (KYON: 27.36° N; 88.714° E) were set up very close to the surface trace of the GTF (Fig. 1.3). Delo Hill (DELO: 27.09° N; 88.50° E), Mungpu (MUNG: 26.98° N; 88.4° E), Namchi (NAMC: 27.16° N; 88.32° E) and Panthang (GBSK: 27.365° N; 88.569° E) were set up away from the GTF fault zone west of the surface trace of the GTF (Fig. 1.3). Nim (NIM: 26.99° N; 88.68° E) was the only station set-up east of the GTF surface trace because of logistical problems and security concerns voiced by the Indian Army over operation of high precision GPS receivers close to international borders. We look at velocities of all the stations in the Darjiling-Sikkim Himalaya relative to the Nim station to work out the slip-rate on the GTF (Table 1.7) during 2000-2004. Mungpu (MUNG: 26.98° N; 88.4° E), which is located west of the GTF surface trace at the same latitude as Nim does not exhibit any significant relative motion with Nim. This indicates that the GTF is locked south of the latitude 27° N (Fig. 1.3). North of 27° N, all stations away from the GTF fault zone viz. Panthang, Delo Hill, Namchi exhibit a southward relative velocity of ~ 4mm/yr with Nim. Stations near the GTF surface trace record higher southward relative velocities north of 27° N (Table 1.7). The northern most station, Kyongnosla, records ~ 7 mm/year (fault-parallel component 4.83 ± 0.86 mm/yr) and Labha records ~ 5 mm/year southward (and approximately fault-parallel) velocity relative to Nim (Table 1.7; Fig. 1.3c). This indicates that there is about 4-5 mm/yr slip on the GTF during 2000-2004 between 27.367° N and 27.07° N. Moreover, as the west block of the fault moves to the south relative to the east block the slip on the GTF is sinistral or left-lateral. This reasoning is supported by ~ 4 mm/yr northward (Table 1.7), fault-parallel motion of Nim (NIMC) station relative to the permanent station at Panthang (Fig. 1.3d). Sinistral strike-slip earthquakes have been recorded near Kyongnosla (Table 1.8); this corroborates the GPS results and additionally indicates that the slip of the GTF is seismogenic.

Table 1.6 : 2000-2004 GPS Convergence Rates in the Darjiling-Sikkim Himalaya.

Baseline	Convergence mm/yr
IISc-LHAS	11.93 ± 0.34
IISc-KYON	06.25 ± 0.84
IISc-GBSK (2003-2004)	03.66 ± 0.57
IISc-LAVA	05.60 ± 0.73
IISc-DELO	03.58 ± 0.35
IISc-NAMC	05.31 ± 0.52
IISc-MUNG	02.86 ± 0.55
IISc-NIMC (2002-2004)	01.81± 1.50
LHAS-KYON	05.98 ± 0.71
LHAS-GBSK (2003-2004)	06.89 ± 0.49
LHAS-LAVA	07.12 ± 0.60
LHAS-DELO	07.57 ± 0.38
LHAS-NAMC	05.59 ± 0.43
LHAS-MUNG	09.51 ± 0.45
LHAS-NIMC (2002-2004)	12.32 ± 1.16

Contd...

KYON-GBSK (2003-2004)	00.72 ± 1.15
KYON-LAVA	01.54 ± 0.59
KYON-DELO	02.11 ± 0.73
KYON-NAMC	-0.82 ± 0.95
KYON-MUNG	03.76 ± 0.78
KYON-NIMC (2002-2004)	06.88 ± 1.12
GBSK-LAVA (2003-2004)	- 1.45 ± 0.57
GBSK-DELO (2003-2004)	- 0.42 ± 0.41
GBSK-NAMC (2003-2004)	- 1.39 ± 0.62
GBSK-MUNG (2003-2004)	02.11 ± 0.49
GBSK-NIMC (2003-2004)	03.40 ± 1.05
LAVA-DELO	00.62 ± 0.92
LAVA-NAMC	- 1.84 ± 0.94
LAVA-MUNG	01.20 ± 0.94
LAVA-NIMC (2002-2004)	04.35 ± 1.35
DELO-NAMC	- 2.25 ± 0.67
DELO-MUNG	01.42 ± 0.56
DELO-NIMC (2002-2004)	01.69 ± 1.42
NAMC-MUNG	02.01 ± 0.42
NAMC-NIMC (2002-2004)	00.32 ± 1.44
MUNG-NIMC (2002-2004)	01.08 ± 1.52

Table 1.7 : Velocities of GPS Stations relative to NIMC and GBSK.

Station (*Permanent Station)	**Velocity(E) (mm/yr) NIMC FIXED**	**Velocity (N) (mm/yr) NIMC FIXED**	**Velocity(E) (mm/yr) GBSK FIXED**	**Velocity (N) (mm/yr) GBSK FIXED**
GBSK*(2003-2004)	-0.93 ± 1.74	-4.02 ± 1.24	-	-
KYON	0.39 ± 1.89	-7.22 ± 1.29	1.32 ± 1.37	-3.20 ± 0.91
NAMC	-2.52 ± 1.72	-4.30 ± 1.23	-1.59 ± 1.13	-0.28 ± 0.83
DELO	-0.18 ± 1.69	-4.60 ± 1.22	0.75 ± 1.08	-0.58 ± 0.82
LAVA	-0.78 ± 1.82	-5.52 ± 1.27	0.15 ± 1.28	-1.50 ± 0.89
NIMC (2002-2004)	-	-	0.93 ± 1.74	4.02 ± 1.24
MUNG	-1.08 ± 1.74	-1.51 ± 1.23	-0.15 ± 1.16	2.51 ± 0.84

Table 1.8 : Focal Mechanisms of moderate strike-slip earthquakes near Kyongnosla *HRVD (Harvard Catalogue) http://www.seismology.harvard.edu/CMTsearch.html. @Dasgupta *et al.*, 2000.

No.	Year	Month	Day	Time	Long.	Lat.	Dep. (km)	Mag.	Focal Mech.	Strike	Dip	Rake
1*	1980	11	19	19:00: 55.9	88.79	27.40	02	6.2	St-slip	209	51	-002
2@	1982	04	05	02:19: 44.6	88.83	27.38	33	5.1	St-slip	206	48	-030

1.3 Implications for Himalayan Deformation

The nature of active tectonics in the Darjiling-Sikkim Himalayas seems to point to many deviations from the recognized models of deformation in the Himalaya. For example, only about 12 mm/yr of convergence is accommodated across the Darjiling-Sikkim and western Bhutan Himalayas as opposed to about ~20mm/yr measured in Central Nepal and postulated for the entire Himalayan arc (Bilham *et al.*, 2001, Bilham *et al.*, 1997, Jouanne *et al.*, 1999, Larson *et al.*, 1999). This suggests that deformation in the Himalayan arc is not continuous and different parts of the arc may be accommodating different amount of convergence. Seismicity patterns along the length of the Himalayan arc also seem to be discontinuous and different parts of the belt seem to be active at different places along the length of the arc (Pandey *et al.*, 1999) suggesting that the Himalayan deformation is segmented into blocks along the length of the mountain belt that tend to move and deform as individual units (Gaur, 1994, Guptasarma, 1996); this is likely to be reflected in detailed and time-averaged GPS measurements from across the length of the Himalayan arc. Some lateral variation is already evident from the published convergence rates measured across the length of the Himalayan arc. The ~20 mm/yr convergence measured in the central part of the Himalayan arc in central Nepal area (Bilham *et al.*, 1997, Jouanne *et al.*, 1999, Larson *et al.*, 1999) appears to be the maximum convergence accommodated in the Himalayan arc (Fig. 1.2); the convergence appears to decrease away from the centre of the arc.

Differences in seismicity and GPS convergence patterns across the GTF suggests that the Himalayan arc may be broken up into different segments along the length of the Himalayas by transport-parallel transverse faults (or lateral and oblique ramps) and each of these segments have a distinct deformation kinematics that may transit gradually or abruptly into the neighbouring segments; transverse faults would result in the latter case. Transverse zones are also observed elsewhere in the Himalayas (Valdiya, 1976, Dubey *et al.*, 2004) and major rivers such as Ganga and Yamuna seem to follow these zones before they enter into the Himalayan foreland. Moreover, the entire Himalayan Mountain front is sinusoidal and segmented into salients and recesses; this suggests lateral variation in deformation kinematics (Marshak, 2004, Macedo and Marshak, 1999) along the length of the Himalayan arc. It follows that the deformation kinematics of each of these blocks needs to be worked out separately and put together to construct the signatures of the overall Himalayan deformation. Given this, cobbling together of geological sections, geological slip-rates, Quaternary deformation signatures and GPS convergence rates from different parts of the Himalayan belt to arrive at global representatives is also not realistic. For example, the results of the INDEPTH Project (Nelson *et al.*, 1996) have been extrapolated to different parts of the Himalayas to prepare representative sections across the Himalaya. We contend that the INDEPTH work is representative of the Himalayan hinterland only north of the Darjiling-Sikkim Himalayas and should not be extrapolated to other parts of the Himalayas especially across transverse zones. Similarly, a present day convergence rate of ~20 mm/year has been measured in the Nepal Himalayas and extrapolated to other parts of the Himalayan arc (Bilham *et al.*, 2001). The rates measured in the Nepal Himalayas are representative of only the central Nepal Himalaya and not necessarily everywhere in the Himalayan arc. Moreover, we need to have a lot more GPS data spread over much longer time-spans to arrive at representative averages. Given all this, transverse zones may have a very significant role in the kinematics of Himalayan deformation and must be, therefore, studied and better understood before global models of Himalayan deformation can become realistic and reliable. These observations suggest that a fresh look at our ideas on the Himalayan deformation is required for there appears to be more complexity in the Himalayan deformation than postulated by the current models.

Acknowledgements

The authors acknowledge Grant # ESS/23/VES/134/2001 from the Department of Science and Technology, India for this work. P. Dileep Kumar and M. B. Ananda were involved with data collection at Lava, Kyongnosla and Namchi in 2000. Y. K. Rai and A. P. Krishna are acknowledged for help with logistics at the Kyongnosla and Namchi stations. Souvik Bannerjee assisted with data collection during the 2004 campaign. Critical reviews from two anonymous reviewers improved the quality of the manuscript. We also thank Parthasarathi Ghosh, the editor of this volume, for his help in publishing this manuscript.

Bibliography

Acharyya, S.K. (1971). Structure and Stratigraphy of the Darjiling Frontal Zone, Eastern Himalaya, *Geological Survey of India Miscellaneous Publications,* 24, pp. 71-90.

Acharyya, S.K. (1976). On the nature of the Main Boundary Fault in the Darjiling Sub-Himalaya, *Geological Survey of India Miscellaneous Publications,* 24, pp. 395-408.

Acharyya, S.K. (1980). Structural framework and tectonic evolution of the eastern Himalaya, *Journal of Himalayan Geology,* 10, pp. 412-439.

Acharyya, S.K. (1994). The Cenozoic Foreland Basin and tectonics of the eastern sub-Himalaya: problems and prospects, *Himalayan Geology,* 15, pp. 3-21.

Altamimi, Z., Sillard, P., Boucher, C. (2002). ITRF2000: A new release of the International Terrestrial Reference frame for earth science applications, *Journal of Geophysical Research Solid Earth,* 107(B10), pp. 2214.

Apel, E., Bürgmann, R., Bannerjee, P., and Nagarajan, B. (2006). Geodetically constrained Indian Plate motion and implications for plate boundary deformation, *Eos Transactions,* AGU, 85(52), Suppl., pp. T51B-1524.

Banerjee, P., and Burgmann, R. (2002). Convergence across the northwest Himalaya from GPS measurements, *Geophysical Research Letters,* 29, pp. 301-304.

Basak, K., and Mukul, M. (2000). Deformation Mechanisms in the South Kalijhora Thrust and Thrust Sheet in the Darjiling Himalayan Fold-and-Thrust Belt, West Bengal, India, *Indian Journal of Geology,* 72, pp.143-152.

Bettinelli, P., Avouac, J., Flouzat, M., Jouanne, F., Bollinger, L., Willis, P., Chitrakar, G. R. (2006). Plate motion of India and Interseismic strain in the Nepal Himalaya from GPS and DORIS measurements, *Journal of Geodesy,* DOI:10.1007/s00190-006-0030-3.

Bilham, R., and Gaur, V. K. (2000). Geodetic contributions to the study of seismotectonics in India, *Current Science,* 79 (9), pp.1259-1269.

Bilham, R., Gaur, V. K., Molnar, P. (2001). Himalayan Seismic Hazard, *Science,* 293, pp. 1442-1444.

Bilham, R., Larson, K., Freymueller, J., and Project Idylhim members, (1997). GPS measurements of present-day convergence across the Nepal Himalaya, *Nature*, 386, pp. 61-64.

Dasgupta, S., Pande, P., Ganguly, D., Iqbal, Z., Sanyal, K., Venkatraman, N. V., Dasgupta, S., Sural, B., Harendranath, L., Mazumdar, K., Sanyal, S., Roy, A., Das, L. K., Misra, P. S., Gupta, H. (2000). Seismotectonic Atlas of India and its Environs, in. P. L. Narula, S. K. Acharyya, and J. Banerjee (eds.), *Geological Survey of India*, pp. 87.

De, R. (1996). A microearthquake survey in the Himalayan Foredeep Region, North Bengal area, *Journal of Himalayan Geology*, 17, pp. 71-79.

De, R. (2000). A microearthquake survey at the main boundary thrust in Sikkim Himalaya, *Journal of Geophysics*, XXI, pp. 77-84.

De, R., and Kayal, J. R. (2003). Seismotectonic Model of the Sikkim Himalaya: Constraint from Microearthquake Surveys, *Bulletin of the Seismological Society of America*, 93, pp.1395-1400.

De, R., and Kayal, J. R. (2004). Seismic activity at the MCT in the Sikkim Himalaya, *Tectonophysics*, 386, pp. 243-248.

Dubey, A. K., Bhakuni, S. S., and Selokar, A. D. (2004). Structural evolution of the Kangra recess, Himachal Himalaya: a model based on magnetic and petrofabric strains, *Journal of Asian Earth Sciences*, 24, pp. 245-258.

Gansser, A. (1983). Geology of the Bhutan Himalaya. Birkhauser Verlag Basel, pp. 181.

Gaur, V. K. (1994). Evaluation of seismic hazard in India towards minimizing earthquake risk. *Current Science*, 67, pp. 324-335.

Guptasarma, D. (1996). Is the Seismic Risk similar everywhere along the Himalayan Collision Belt? *Himalayan Geology*, 17, pp. 1-9.

Hoffmann-Wellenhof, B., Lichtenegger, H., Collins, J. (1997). GPS Theory and Practice: Springer-Verlag Wein, New York. pp. 389.

Jade, S. (2004). Estimates of plate velocity and crustal deformation in the Indian subcontinent using GPS geodesy, *Current Science*, 86, pp. 1443-1448.

Jade, S., Bhat, B. C., Bendick, R., Ananda, M. B., Dileep Kumar, P., Molnar, P., Gaur, V. K. (2004). GPS measurements from the Ladakh Himalaya, India: Preliminary tests of plate-like or continuous deformation in Tibet, *Geological Society of America Bulletin*, 116 (11), pp. 1385 1391.

Jade, S., Mukul, M., Bhattacharyya, A. K., Vijayan, M. S. M., Saigeetha, J., Kumar, Ashok., Tiwari, R. P., Kumar, Arun, Kalita, S., Sahu, S. C., Krishna, A. P., Gupta, S. S., Murthy, M. V. R. L., Gaur, V. K. (2007). Estimates of interseismic deformation in Northeast India from GPS Measurements. *Earth and Planetary Science Letters*, 263, pp. 221-234.

Jouanne, F., Mugnier, J., Pandey, M., Gamond, J. F., LeFort, P., Serrurier, L., Vigny, C., Auovac, J. P. (1999). Oblique convergence in the Himalayas of western Nepal deduced from preliminary results of GPS measurements, *Geophysical Research Letters,* 26, pp. 1933-1936.

Kayal, J. R. (2001). Microearthquake activity in some parts of the Himalaya and the tectonic model, *Tectonophysics,* 339, pp. 331-351.

King, R. W., and Bock, Y. (2000), Documentation of the GAMIT GPS analysis software, Massachusetts Institute of Technology, Cambridge, Scripps Institution of Oceanography University of California San Diego.

Larson, K. M., Burgmann, R., Bilham, R. and Freymueller, J. T. (1999). Kinematics of the India Asia collision zone from GPS measurements, *Journal of Geophysical Research,* 104, pp. 1077-1093.

Marshak, S. (2004). Arcs, Oroclines, Salients, and Syntaxes - The origin of map-view curvature in fold-thrust belts, *American Association of Petroleum Geologists Memoir,* 82, pp. 131-156.

Macedo, J., and Marshak, S. (1999). The geometry of fold-thrust belt salients, *Geological Society of America Bulletin,* 111, pp. 1808-1822.

Mukul, M. (2005). Continental deformation and Global Positioning System based Geodesy: *Himalayan Geology,* 26 (1), pp. 193-98.

Mukul, M. (2000). The geometry and kinematics of the Main Boundary Thrust and related neotectonics in the Darjiling Himalayan fold-and-thrust belt, West Bengal, India, *Journal of Structural Geology 22,* pp. 1261-1283.

Mukul, M., and Matin, A. (2005). Tectonics of the Himalayan Mountain Front, Darjiling Himalaya, India. *Abstracts, 2005 Asia Oceania Geosciences Society 2nd Annual Meeting,* Singapore, June 20-24, 2005, p. 280.

Mukul, M., Jaiswal, M., Singhvi, A. K. (2007). Timing of recent out-of-sequence active deformation in the frontal Himalayan wedge: Insights from the Darjiling sub- Himalaya, India, *Geology,* 35 (11), pp. 999-1002.

Nakata, T. (1989). Active faults of the Himalaya of India and Nepal, *Geological Society of America Special Paper,* 232, pp. 243-264.

Nath, S. K., Vyas, M., Pal, I., Sengupta, P. (2005). A Seismic Hazard Scenario in the Sikkim Himalaya from Seismotectonics, Spectral Amplification, Source Parametrization and Spectral Attenuation Laws using Strong Motion Seismometry. *Journal of Geophysical Research – Solid Earth,* 110, B01301, doi:10.1029/2004JB003199.

Nelson, K.D., and twenty six others, (1996). Partially Molten Middle Crust Beneath Southern Tibet: Synthesis of Project INDEPTH Results, *Science,* 274, pp. 1684-1688.

Ni, J. and Barazangi, M. (1984). Seismotectonics of the Himalayan Collision Zone: Geometry of the underthrusting Indian plate beneath the Himalaya, *Journal of Geophysical Research,* 89 (B2), pp. 1147-1163.

Pandey, M. R., Tandulkar, R. P., Avouac, J. P., Vergne, J., Heritier. Th. (1999). Seismotectonics of the Nepal Himalaya from a local seismic network, *Journal of Asian Earth Sciences*, 17, pp.703-712.

Raina, V. K. (1976). The Rangit Tectonic Window - stratigraphy, structure & tectonic interpretation and its bearing on the regional stratigraphy, *Geological Survey of India Miscellaneous Publication*, 41, pp. 36-42.

Ray, S. K. (2000). Culmination Zones in Eastern Himalaya, *Geological Survey of India Miscellaneous Publications*, 55, pp. 85-94.

Schwan, W. (1980). Shortening structures in eastern and north-western Himalayan rocks, in P. S. Saklani, (ed.) Today and Tomorrow Printers and Publishers, New Delhi.

Socquet, A.,Vigny, C., Rooke, N. C., Simons, W., Rangin, C., Ambrosius, B. (2006). India and Sunda plates motion and deformation along their boundary in Myanmar determined by GPS, *Journal of Geophysical Research*, 111, B05406, doi: 10.1029/2005JB003877.

Vingy, C., Socquet, A., Rangin, C., Chamot-Rooke, N., Pubellier, M., Boudin, M.-N., Bertrand, G., Becker, M. (2003). Present-day crustal deformation around Sagaing fault, Myanmar. *Journal of Geophysical Research*, 108 (B11), 2533, doi:10.1029/2002JB001999.

Valdiya, K. S. (1976). Himalayan Transverse Faults and Folds and their Parallelism with Subsurface Structures of North Indian Plains, *Tectonophysics* 32, pp. 353-386.

CHAPTER 2

Preliminary Results of a Study of Crustal Deformation in the Himalayan Frontal Zone in North Bengal using GPS Geodesy

Mallika Mullick, Dhruba Mukhopadhyay and Bikash C. Poodar

Abstract

Ten GPS stations have been set up in campaign mode in North Bengal in the Himalayan foothills and frontal zone, of which 7 have been reoccupied. Data from these 7 GPS stations over the period December, 2005 - January, 2007 have been processed using GAMIT/GLOBK suite of software and the velocities of these stations have been estimated. Since the data obtained so far cover a span of one year only, the conclusions derived from them are to be considered as preliminary. The stations are found to be moving at the rate of ~ 40-55 mm/yr in ITRF05 reference frame towards NNE. The baseline between LHAS (Lhasa) and IISC (Bangalore) is estimated to be shortened by ~10.2 mm over the period of our measurement. The southernmost station in our study, NBUV, at Siliguri is more or less collinear with LHAS-IISC baseline. The shortening of NBUV-IISC line is estimated to be ~ 3.8 mm over this period and that of NBUV-LHAS line is ~ 6.4 mm for the same period. These GPS stations were set up across the parallel set of previously marked fault scarps named Matiali (MT), Chalsa (CL) and Baradighi (BD) faults in North Bengal between longitudes 88^0 E and 89^0 E. The results from the GPS measurements show that stations to the south of the BD fault are moving at a faster rate than those to the north of MT fault.

The baseline between stations AMBK (Ambiok) and NBUV has shortened by ~3 mm over the period of our observation. These two are the northernmost and southernmost stations respectively with respect to the three E-W running parallel faults and thus the shortening of baseline AMBK-NBUV is due to movements on the set of three faults taken together. A N-S to NNE-SSW running line of discontinuity between 88.5^0 E and 89^0 E longitudes separates the velocity vectors of the set of stations SAMS, JITI, PMSM, BRDH to the east of the line and the set of stations NBUV, BGKT, AMBK to the west. The line of demarcation of the two blocks containing the two sets of stations appears to coincide with the Gish transverse zone (GTZ).

2.1 Introduction

The Himalayan mountain system is the result of an on-going collision between the Indian plate and the Eurasian plate that started about 40-50 Ma ago (Molnar, 1986). The collision has led to a complex deformation process along the Himalayan arc involving the two plates. The convergence process is however not the same all along the 2500 km stretch of the Himalayas. In the Ladakh Himalayas the convergence rate is 15+ 2 mm/yr (Banerjee and Burgmann, 2002) and in the Sikkim Himalayas it has been measured to be 10-12 mm/yr by Jade, (2004). GPS measurements in the Nepal Himalayas show a mean convergence rate across Central and Eastern Nepal to be 19±2.5 mm/yr (Bilham *et al.,* 1997).

The structural architecture of the Himalayas is controlled by a number of stacked, complexly deformed thrust sheets (Gansser, 1964, Hodges, 2000 and references therein). The principal regional thrust planes are, from north to south, the Main Central Thrust (MCT), the Main Boundary Thrust (MBT), and the Main Frontal Thrust or Himalayan Frontal Thrust (MFT or HFT). All of them are thought to be splays from a main basal detachment fault or Himalayan Sole Thrust (HST) (Hodges *et al.,* 2001, Hodges *et al.,* 2004). In the commonly accepted view of structural development of the Himalayas, faulting propagated from north to south, with the MCT being active in early Miocene time, being succeeded by the Pliocene MBT and by the currently active MFT (Gansser 1964, Hodges, 2000, Le Fort, 1975, Ni and Barazangi, 1984). In this model no active, surface-breaking thrust occurs north of the trace of MFT (Cattin and Avouac, 2000). In an alternative model active faults include both the reactivated segments of MCT system and the new and reactivated structures of the Lesser Himalayan Duplex (Hodges *et al.*, 2001, Hodges *et al.*, 2004). Lave and Avouac (2001) concluded on the basis of geomorphologic evidence that the MFT in central Nepal experienced ca. 2 cm/yr shortening averaged over the past 10000 years. This rate is comparable to the modern rate of convergence across this sector of the Himalayas (Bilham *et al.*, 1997, Larson *et al.*, 1999), suggesting that the shortening is entirely accommodated by slip on MFT and HST, without being partitioned into MBT or MCT.

The generalized pattern of MBT and MCT in the eastern Himalayas is shown in Fig. 2.1. In the foothills of North Bengal between latitudes 26.5^{0}N and 27^{0}N, Nakata (1989) considered that some E-W scarps within the Quaternary sediments mark the traces of "active" faults (Fig. 2.1). Prominent among these are the Chalsa and Matiali scarps (Fig. 2.2). These are regarded by Nakata to mark the traces of the MFT and MBT respectively. However, the kinematics and the timing of movement on these faults were not discussed in details.

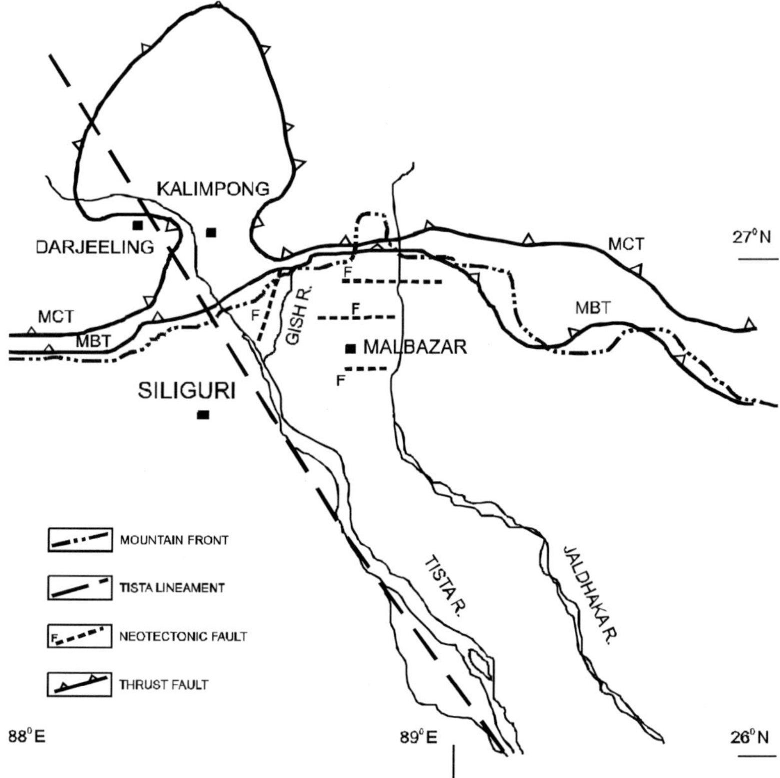

Fig. 2.1 : A generalized tectonic map of the present area of investigation in North Bengal (modified from Seismotectonic Atlas of India and its Environs, Geological Survey of India, Dasgupta *et al.*, 2000).

Recently Mukul and Matin (2005) have reported the existence of a transverse structure along the Gish river valley, referred to as the Gish transverse zone (GTZ), at high angles to the trend of the Himalayan belt. They have pointed out the difference of the structural architecture east and west of the GTZ and have commented that the mountain front east of the GTZ is characterized by asymmetric river terraces that have been raised by ~ 400 m relative to the present level of the Chel River. Such terraces are missing west of the GTZ. Mukul *et al.* (2005) have also reported the results of GPS measurements of several campaign mode stations in the Darjeeling-Sikkim Himalaya (DSH). The convergence rate between Lhasa (LHAS) and Bangalore (IISC) is estimated to be ~10mm/yr, and only about half of this convergence is reported to be taken up in the DSH. For a campaign mode station at Delo Hill (DELO: 27.09 N; 88.50 E), the convergence between LHAS and DELO during 2000-2003 was 5.67 ± 1.68 mm/yr. The measurements reveal a statistically insignificant convergence of 1.29 ± 1.76 mm/yr between IISC & DELO during 2000- 2003. For a station a little to the south of DELO at Mungpu (MUNG: 26. 98 N; 88. 40 E) the IISC-MUNG convergence over 2000-2003 was statistically insignificant 1.66 ± 1.94 mm/yr whereas LHAS-MUNG convergence over the same period was 10.1 ± 1.87 mm/yr. The 2000-2003 DELO-MUNG convergence was measured to be 3.83 ± 2.07 mm/yr. From these results they concluded that the frontal part of the DSH is not accommodating much convergence and most of the 4-5 mm/year convergence is being taken up between 27°N and 27.5°N latitudes. De and Kayal (2004), on the basis of micro-earthquake study, inferred right-lateral strike slip on a nearly vertical NNE-SSW trending transverse fault. They correlated it with the surface trace of a NNE-SSW trending segment of MCT in this region. However, the inferred fault may well be the northern continuation of the GTZ.

The present work was undertaken to monitor the active crustal deformation in the Himalayan foothills of North Bengal using GPS geodesy, and specifically to test if any movement is currently taking place on the active E-W trending faults inferred by Nakata (1989). It was also our objective to find out how much of the current total shortening in the Eastern Himalayas is accommodated by movement along faults in the Himalayan foothills.

2.2 GPS Data Acquisition and Processing

A network of 10 GPS stations (Fig. 2.2) was set up around and across the Matiali (MT), Chalsa (CL) and Baradighi (BD) faults mapped by Nakata (1989). Amongst these, 7 have been reoccupied for estimation of site velocity. The GPS data were collected in campaign mode using Trimble 5700 dual frequency receiver and Zephyr Geodetic antenna set. At each station the data were collected continuously for 4-5 days. The stations were set up during December, 2005, January and March, 2006. They were reoccupied during December, 2006 and January and March, 2007. The instrument sets were carefully guarded for the entire epoch of measurement in order to ensure their stability. All the stations were set up in the premises of some important institutions so that the monuments along with the markings made on them could be well-guarded and be easily retrieved after intervals of one year. As an example the time series of station NBUV is shown over the two epochs at an interval of one year in Figs. 2.3a and 2.3b.

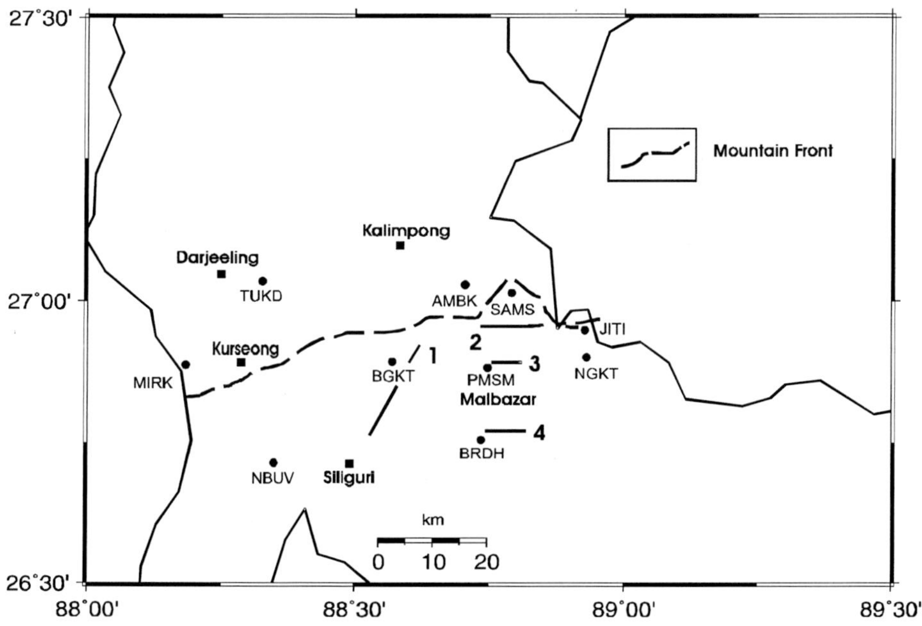

Fig. 2.2 : Map showing location of 10 campaign mode GPS stations in North Bengal and the following fault traces. 1 - Gish Transverse Zone, 2 - Matiali fault, 3 - Chalsa fault, 4 - Baradighi fault.

The data processing was carried out using the GAMIT/GLOBK suite of software (Dong et al, 1998) in the computer lab at CSME, Kolkata. GAMIT is a comprehensive GPS analysis package developed at MIT, USA and Scripps Institution of Oceanography, USA (Herring *et al.*, 2006) for estimation of 3-D relative positions of ground stations

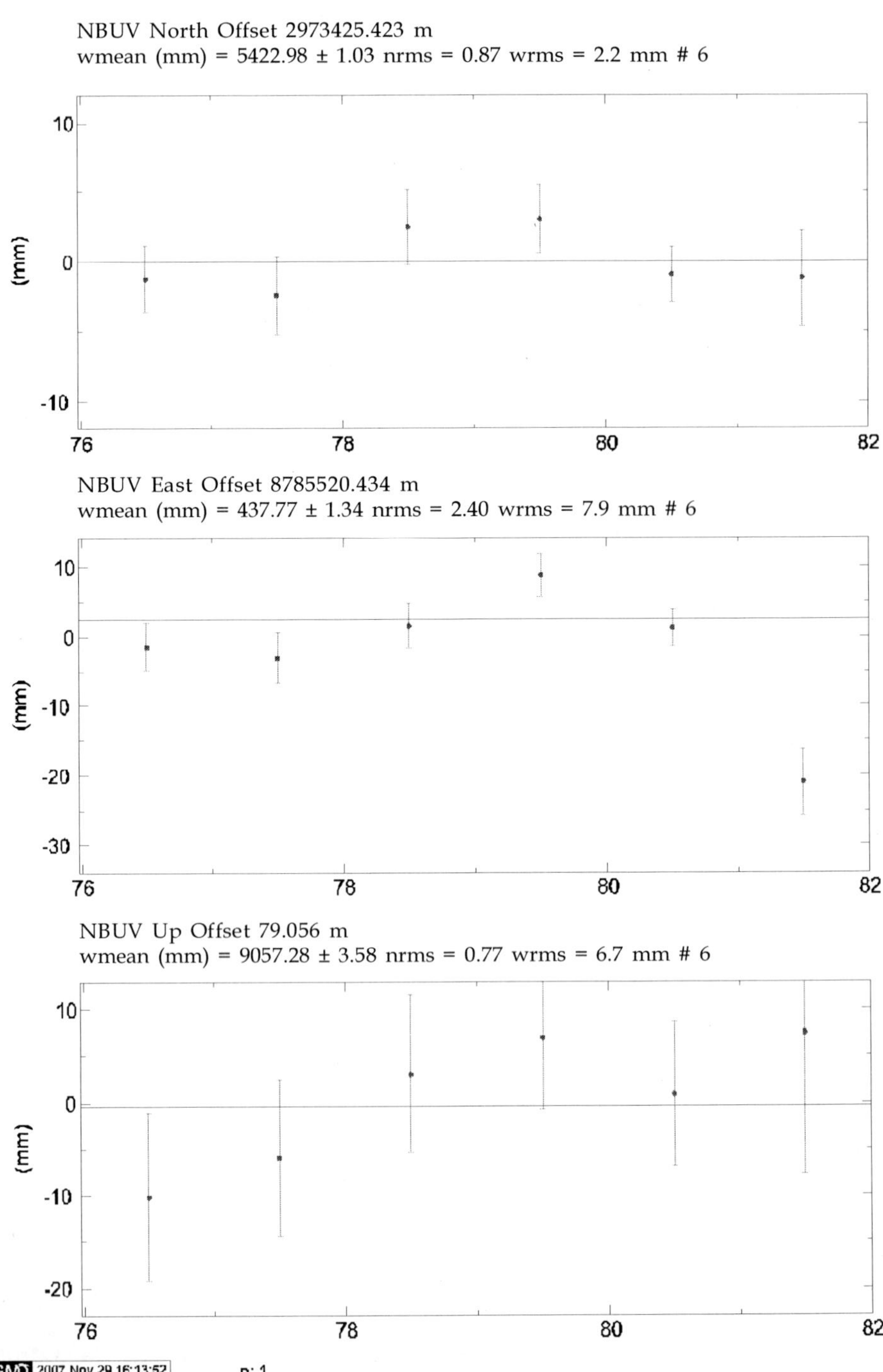

Fig. 2.3a : Time series of station NBUV during first epoch of measurement.

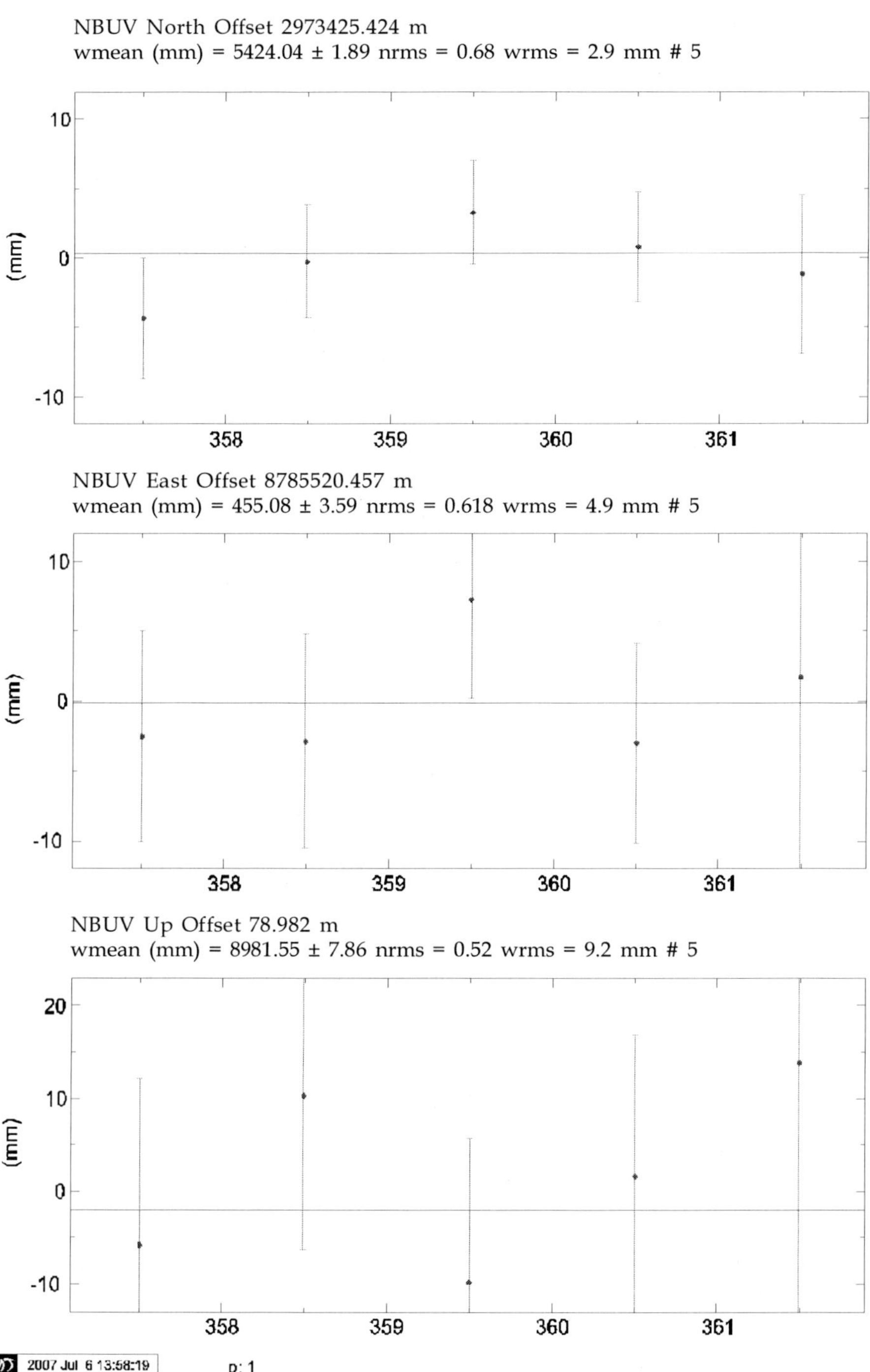

Fig. 2.3b : Time series of station NBUV during second epoch of measurement.

and satellite orbits. The phase and pseudo-range observations recorded in the GPS receivers are converted to RINEX format. The RINEX files of our local stations, along with those from IGS (International GNSS Service) stations–IISC, POL2, KIT3, KUNM, HYDE, ULAB, URUM, PIMO, GUAO, satellite broadcast ephemeris data and data about orbital initial conditions for all satellites are given as input to GAMIT. The primary output of GAMIT are 'quasi-observables' or a loosely constrained solution of co-variances and estimate of station coordinates, satellite orbits, tropospheric zenith delays, earth orientation parameters and phase ambiguities of the carrier waves estimated daily and independently by the weighted least square technique. This output is passed on to GLOBK, a Kalman filter (Herring *et al.*, 2006), in combination with global solutions from IGS stations to estimate station positions, velocities, orbital and Earth-rotation parameters in a well-defined reference frame. While running GLOBK we have used the data from about 25-30 IGS stations with well-determined, *a priori* coordinates for defining the reference frame with a good postfit rms values of 0.7-2 mm/yr.

2.3 GPS Results

We report here the preliminary estimates of the velocities of 7 GPS stations around and across CL, MT and BD faults determined on the basis of observations taken over a period of 1-year (Fig. 2.4, Table 2.1). Since the data are collected over a short time span, all the tectonic implications are to be treated as tentative and not as an established fact. More precise and accurate quantification would be obtained after measurements over a longer duration (>3 years) become available. Since velocity of 7 GPS stations along with that of IGS stations LHAS (Lhasa) and IISC (Bangalore) cannot be shown clearly in one map, as an example, we show the velocity of station NBUV (North Bengal University, Siliguri), IISC and LHAS in ITRF05 reference frame in Fig. 2.5. Interestingly, the stations IISC, NBUV and LHAS are almost collinear; NBUV is situated within the Quaternaries of the Himalayan frontal zone. NBUV is moving at the rate of ~ 48.6 mm/yr along NNE. Figure 2.6 shows the velocity of NBUV and LHAS with respect to IISC. The IISC-LHAS baseline is measured to be shortened by ~ 10.2 mm over the period of December, 2005 - January, 2007. This value is in agreement with the rate obtained by Jade (2004). This shortening is quite different from the value of convergence rate in the Ladakh-Garhwal Himalayas and in Nepal Himalayas. In Ladakh-Garhwal region the convergence is estimated to be ~15+2 mm/yr (Banerjee and Burgmann, 2002) and ~20mm/yr in Nepal (Bilham *et al.*, 1997). The velocity of NBUV is ~3.8 mm/yr with respect to IISC and ~6.4 mm/yr with respect to LHAS. These estimates imply that the NBUV-IISC baseline has shortened by ~ 3.8 mm and the NBUV-LHAS baseline has shortened by ~ 6.4 mm over the period of our measurement. These preliminary estimates indicate the major part of the IISC-LHAS shortening is accommodated within the Himalayan zone.

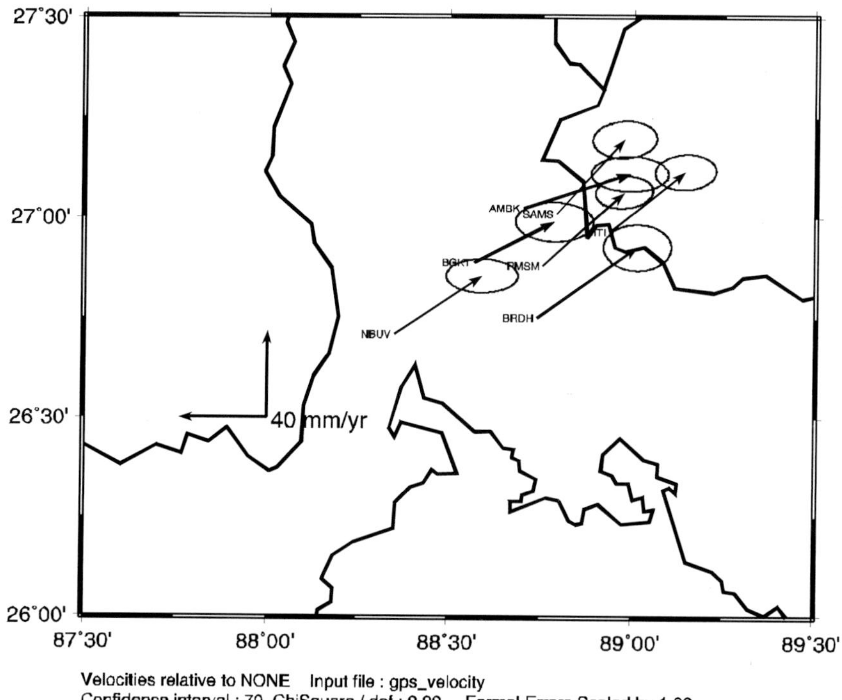

Fig. 2.4 : Velocity of 7 campaign mode GPS stations in North Bengal estimated in ITRF05 reference frame plotted along with their error ellipses.

A second conclusion is suggested from a consideration of the velocity vectors of the 7 stations (Fig. 2.4) estimated in ITRF05 reference frame. The stations to the south of the BD fault are seen to be moving at a faster rate than those to the north of the CL and MT faults. Stations BRDH (Baradighi village) and NBUV located south of the BD trace are moving with velocities ~54.8 mm/yr and ~48.6 mm/yr respectively, while stations SAMS (Samsing forest village) and JITI (Jiti tea garden) located to the north of the MT fault are moving with velocity ~45.9

mm/yr and ~44.1 mm/yr respectively. AMBK (Ambiok) is the northern most station of our network in North Bengal across the set of three faults, CL, MT and BD. The AMBK-NBUV baseline has shortened by ~3 mm over the period of our observation. AMBK is nearly collinear with IISC-NBUV-LHAS baseline. The above estimate along with the observation of the velocity of NBUV with respect to LHAS (~ 6.4 mm/yr), give us an indication that about half of the total shortening in the Himalayan mountain belt is accommodated in the foothills and frontal zone.

Table 2.1 : GPS derived velocities of campaign mode and IGS stations in ITRF05 reference frame, their locations and azimuth.

Location of stations	Station code	Remarks	Longi-tude (deg)	Lati - tude (deg)	Velocity (mm/yr)	Azimuth (deg)
Lhasa	LHAS	IGS Station	91.104	29.657	43.76±4.2	80.91
Bangalore	IISC	IGS Station	77.570	13.021	53.1±4.0	53.244
Samsing forest village	SAMS	Campaign Station	88.791	27.010	45.9±9.8	41.642
Jiti tea garden	JITI	Campaign Station	88.936	26.965	44.1±9.3	50.926
Baradighi village	BRDH	Campaign Station	88.738	26.751	54.8±10.9	54.379
North Bengal University, Siliguri	NBUV	Campaign Station	88.350	26.711	48.6±10.5	55.966
Ambiok	AMBK	Campaign Station	88.702	27.024	50.3±11.3	71.446
Malbazar	PMSM	Campaign Station	88.753	26.881	49.9±8.6	47.229
Bagrakot	BGKT	Campaign Station	88.569	26.889	41.4±11.6	62.400

The baseline between SAMS, north of MT fault and PMSM (Malbazar), south of CL fault has shortened by ~2 mm over the period of our observation. This gives an estimate of the shortening along N-S direction that occurs across the CL and MT faults taken together. The perpendicular distance between the faults CL and MT faults is less than 10 km and hence no GPS station was set up in that gap to independently monitor the movement on CL and MT faults.

The velocity vectors at the different stations (Fig. 2.4) tend to suggest another inference that a N-S / NNE-SSW trending line of discontinuity exists between longitudes 88.5^0 E and 89^0 E. The set of stations NBUV, AMBK and BGKT west of the discontinuity are moving more eastward than the eastern set of SAMS, JITI, and PMSM. The magnitudes of velocities of the two sets show slight difference. The line of discontinuity approximately coincides with the Gish transverse zone (GTZ) recognized by Mukul and Matin (2005) and Dasgupta et al. (2000) in Seismotectonic Atlas of India. The study of De and Kayal (2004) on micro-earthquakes also suggests active seismogenic faulting along such a transverse zone. However the large error ellipses of the velocity vectors (Fig. 2.4) restrict us to deduce any definite conclusion about the presence of any transverse fault. More accurate tectonic interpretation is expected after the GPS data from a bigger time span (>3 years) is obtained.

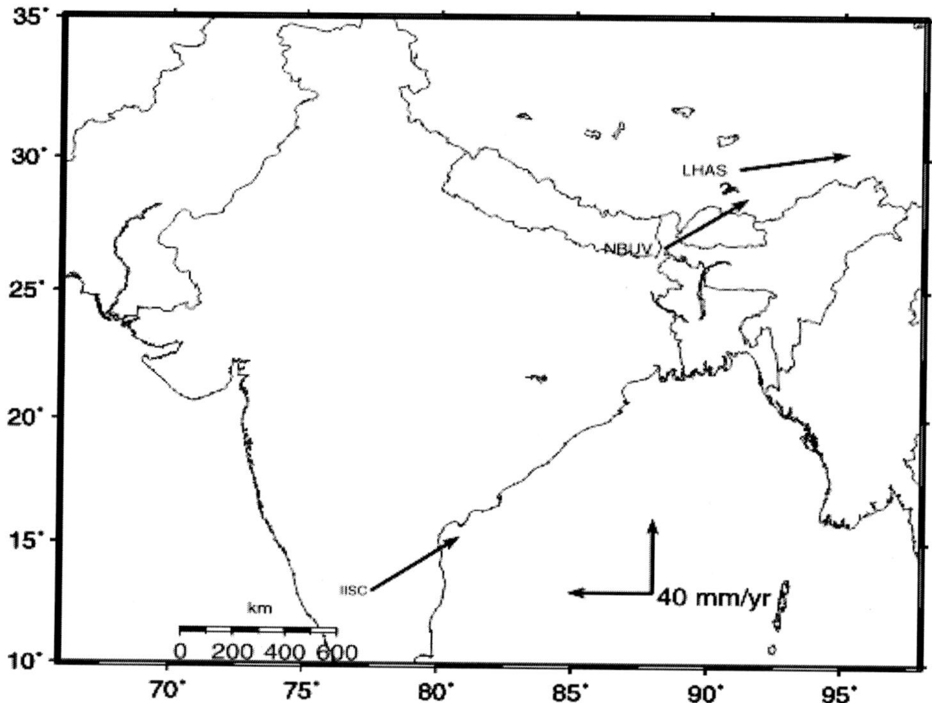

Fig. 2.5 : Velocity of station NBUV and IGS stations LHAS (Lhasa) and IISC (Bangalore) estimated in ITRF05 reference frame plotted on a map of India and Eurasia.

2.4 Conclusions

The present research work has enabled us to arrive at few conclusions. These conclusions are based on GPS data from a short time span of one year and hence should be treated to be absolutely preliminary.

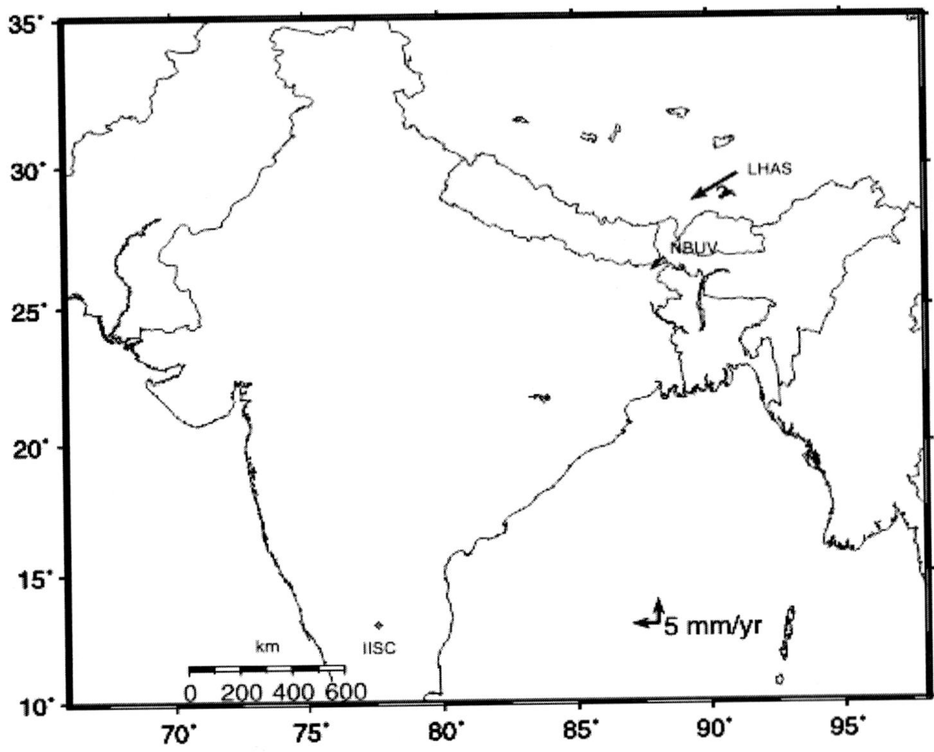

Fig. 2.6 : Velocity of station NBUV and IGS station LHAS (Lhasa) relative to IISC (Bangalore) plotted on a map of India and Eurasia.

a) The LHAS-IISC baseline has shortened by ~10.2 mm which is consistent with the results reported by Jade (2004) and Mukul and Matin (2005).

b) The shortening rate between Himalayan frontal foothills and LHAS is ~6.4 mm over the period of one year.

c) AMBK-NBUV baseline has shortened by ~3mm. Hence about half of the shortening mentioned above is accommodated within the zone with three E-W faults Chalsa, Matiali and Baradighi mapped by Nakata (1989) in the Himalayan foothills and frontal zone in North Bengal. The remaining part (~4 mm) is accommodated by shortening in the lesser and higher Himalayas. The shortening by ~2 mm of SAMS-PMSM baseline traversing the Chalsa and Matiali faults gives an estimate of shortening accommodated by movements on the Matiali and Chalsa faults taken together.

d) The above results support the model that segments of the MCT system and of the thrusts in the Lesser Himalayan Duplex are also currently active (Hodges, 2001, Hodges *et al.*, 2004).

e) The GPS stations to the south of the Baradighi fault are moving faster than the stations to the north of the Matiali fault.

f) A transverse line of discontinuity in the velocity vectors is observed between longitudes 88.5^0 E and 89^0 E. The line of discontinuity appears to represent the Gish transverse zone (GTZ) inferred by Mukul and Matin (2005) and the Seismotectonic map by Dasgupta *et al.*, (2000) (Fig. 2.1).

Acknowledgements

The present research work was carried out under a project sponsored by the Department of Science and Technology, Government of India. The authors are grateful to Dr Paramesh Banerjee of Wadia Institute of Himalayan Geology for teaching us the intricacies of GPS data processing and for his valuable suggestions at all stages of the work. Thanks are also due to Dr Robert King of MIT, USA for his help and suggestions during GPS data processing. We are grateful to Dr Swadesh Kumar Basu for providing some important information about the geology of North Bengal. Authors are thankful to Dr Parthasarathi Ghosh of ISI,Kolkata for helping them in improving the quality of figures. Authors would also like to thank Mr. Subir Chakraborty, Mr. Partha Guha Thakurta and Mr. Narendra B. Singh of CSME for their assistance during the field work.

Bibliography

Banerjee, P. and Burgmann, R. (2002). Convergence across the northwest Himalaya from GPS measurements, *Geophysical Research. Letters*, 29(13), pp. 30-1 - 30-4.

Bilham, R., Larson, K., Freymuller, J., and others (1997). GPS measurements of present-day convergence across the Nepal Himalaya, *Nature*, 386, pp. 61-64.

Cattin, R. and. Avouac, J. P. (2000). Modeling mountain building and the seismic cycle in the Himalaya of Nepal, *Journal of Geophysical Research*, 105, pp. 13389-13407.

Chen, Z., B. C. Burchfiel, Y. Liu, R. W. King, L. H. Royden, W. Tang, E. Wang, J. Zhao, and X. Zhang (2000). Global Positioning System measurements from eastern Tibet and their implications for India/Eurasia intercontinental deformation, *Journal of Geophysical Research*, 105, pp.163215-16227.

Dasgupta, S., Pande, P., Ganguly, D., Iqbal. Z., Sanyal, K., Venkatraman, N.V., Dasgupta, S., Sural, B., Harendranath, L., Mazumdar, K., Sanyal, S., Roy, A., Das, L.K., Misra, Gupta, H. (2000). Seismotectonic Atlas of India and Its Environs, P.L. Narula, S.K. Acharyya, J. Banerjee (Eds.). Geological Survey of India, 87 p.

De, R., and Kayal, J.R. (2004). Seismic activity at the MCT in Sikkim Himalaya, *Tectonophysics*, 386, pp. 243– 248.

Dong, D.N, Herring, T. A. and King, R. W. (1998). Estimating regional deformation from a combination of space and terrestrial geodetic data, *Journal of Geodesy*, 72, pp. 200-214.

Gansser, A. (1964). Geology of the Himalayas. Wiley Interscience, London, 1964, 289 p.

Herring, T.A, King, R.W., McClusky, S.C. (2006) .GAMIT Reference Manual, GPS Analysis at MIT, Release 10.3. Department of Earth, Atmospheric and Planetary Sciences, Massachusetts Institute of Technology. Website : www-gpsg.mit.edu

Herring, T.A, King, R.W., McClusky, S.C. (2006). GLOBK Reference Manual, Global Kalman filter VLBI and GPS analysis program, Release 10.3. Department of Earth, Atmospheric and Planetary Sciences, Massachusetts Institute of Technology. Website : www-gpsg.mit.edu

Hodges, K.V. (2000). Tectonics of the Himalaya and southern Tibet from two perspectives, *Geological Society of America Bulletin*, 112, pp. 324-350.

Hodges, K. V., Hurtado, J.M., Whipple, K.X. (2001). Southward extrusion of Tibetan crust and its effect on Himalayan tectonics, *Tectonics*, 20, pp. 799-809.

Hodges, K. V., Wobus, C., Ruhl, K., Schildgen, T., Whipple, K. (2004). Quaternary deformation, river steepening, and heavy precipitation at the front of the Higher Himalayan ranges, *Earth Planetary Science Letters*, 220, pp. 379-389.

Jade, S. (2004). Estimates of plate velocity and crustal deformation in the Indian subcontinent using GPS geodesy, *Current Science*, 86, pp. 1443-1448.

Larson, K., Bürgmann, R., Bilham, R., Freymueller, J. (1999). Kinematics of the India-Eurasia collision zone from GPS measurements. *Journal of Geophysical Research*, 104, pp. 1077-1093.

Lave, J., Avouac, J.P. (2001). Fluvial incision and tectonic uplift across the Himalayas of central Nepal, *Journal of Geophysical Research*, 106, pp. 26561-26591.

LeFort, P. (1975). Himalayas: The collided range: Present knowledge of the continental arc, *American Journal of Science*, 275-A, pp. 1-44.

Molnar, P. (1986) .The geologic history and structure of the Himalaya, *American Scientist*, 74, pp. 144-154.

Mukul, M. and Matin, A. (2005). Tectonics of the Himalayan Mountain Front, Darjiling Himalayas, India. Annual Report of Centre of Mathematical Modeling and Computer Simulation, Bangalore, 2004-2005, pp. 26. Website: www.cmmacs.ernet.in /cmmacs/ Publications/Annual_rep/2004-05/ chapter2.pdf

Mukul, M., Jade, S., Matin, A., and Vijayan, M. S. M. (2005). Global Positioning System (GPS)-based crustal deformation of the Darjiling-Sikkim Himalaya. Annual Report of Centre of Mathematical Modeling and Computer Simulation, Bangalore, 2004-2005, pp. 26-27. Website: www.cmmacs. ernet.in/cmmacs/ Publications/Annual_rep/2004-05/chapter2.pdf

Nakata, T. (1989). Active faults of the Himalaya of India and Nepal, *Geological Society of America Special Paper*, 232, pp. 243-264.

Ni, J., and Barazangi, M. (1984). Seismotectonics of the Himalayan collision zone: geometry of the under thrusting Indian Plate beneath the Himalaya, *Journal of Geophysical Research*, 89, pp. 1147–1163.

CHAPTER 3

Spatial and Temporal Variations of the Strain Fields in Orogenic Belts : An Analysis Based on Kinematic Models

Atin Kumar Mitra, Manhal Mahmoud
Shamik Sarkar and Nibir Mandal

Abstract

This paper presents a kinematic analysis for explaining the spatial and temporal variations of deformational structures in the Cuddapah Proterozoic orogenic belt, South India. This belt shows virtually undeformed, flat-lying sedimentary rocks in its western margin, whereas strongly deformed and folded rocks in the eastern boundary. The structural variations indicate increasing deformation in the east direction, implying a relative tectonic convergence from the east side. Based on the corner flow theory, we use a viscous model to determine the heterogeneous strain field in the asymmetric convergent zone, and explain varying deformations from the foreland to hinterland. According to this model, the strain field in the hinterland part varies in a pulsating manner with progressive crustal thickening, giving rise to multiple deformational structures in the course of a single continuous tectonic movement. We also studied the strain field in finite element models considering the gravity forces, and obtained similar results. The foreland vergence of deformational structures in the Cuddapah belt indicates a top-to-west horizontal shear during the overall horizontal contraction. Our numerical models show that this horizontal shear component in the foreland region is likely to be associated with a vertical shear component in the hinterland.

3.1 Introduction

Orogenic belts generally contain tectonic structures of multiple generations, indicating complex spatial and temporal evolution of the strain fields. A line of study in structural geology is concerned with geometrical analysis of structural superpositions, such as folds, interference patterns and overprinting of tectonic fabrics (Sutton and Watson, 1955, 1959, Weiss *et al.*, 1955, Ramsay, 1967, Naha and Mohanty, 1988, Ghosh, 1993, Twiss and Moores, 2007). Such studies are primarily aimed at interpreting the orientations of different phases of deformational structures, e.g. folds, foliations, lineations etc. Using kinematic approach, many workers developed models concerning strain analyses and the temporal variations of flow in multiply deformed terrains (Ramsay and Huber, 1987, Macaya *et al.*, 1991, Chattopadhya and Mandal, 2002). However, there have been few attempts for developing a kinematic basis of temporal strain variations in the course of a single orogenic movement. The importance of this issue can be addressed with the help of a planar tectonic foliation that forms parallel to the XY plane of finite strain, but subsequently folds during continued deformation. In case of rotational strain, the foliation may, however, rotate as a passive marker in progressive deformations, and show angular deviations with the rotating XY plane of finite deformation. Albeit, the amount of deviation should be negligibly small (Ghosh, 1982). According to the homogeneous flow model, the foliation is unlikely to undergo shortening, and fold in the same kinematic framework (Ramsay, 1967, Ramberg, 1975, Sanderson, 1979, Ghosh, 1993). Superposition of folds on the foliation thus indicates that orogenic movements involve a temporal variation in the kinematic framework, which needs to be investigated in view of the large-scale dynamics of orogens. Fold belts also show conspicuous variations in the spatial distributions of deformational structures. For example, Fyson (1971) has shown that folds at shallow crustal levels are mostly upright, whereas those at greater depths are recumbent (see Ghosh, 1993 for detailed discussion). The stretching lineation occurs at a high angle to the fold hinges of upright structures, while it is either at a low angle or parallel to hinges of recumbent folds. Field investigations reveal such variations also in the horizontal directions.

Based on observations from the Cuddapah Proterozoic orogenic belt, South India this paper presents an analysis of the spatial and temporal variations in the strain field. In geodynamics the workers have used a variety of continuum models based on different crustal rheology, e.g. Coulomb, elastic-plastic and visco-elastic etc., which investigate the kinematics of orogenic wedges (Davis *et al.*, 1983,

Dahlen, 1984, 1990, Wang and Davis, 1996, Stockmal, 1983, Chapple, 1978, England and Houseman, 1985, Willett, 1992, 1999). A line of these studies involves viscous models, considering thick-skinned crustal tectonics and continuous ductile flow in orogenic belts (Platt 1986, Grujic *et al.*, 1996, Royden 1996, Willett 1999, Rossetti *et al.*, 2000, Beaumont *et al.*, 2001, Jiménez- Munt *et al.*, 2006, Groome *et al.*, 2008). We follow this approach, and employ a viscous model based on the corner flow theory (Batchelor, 1967). This theory has been used by several earlier workers for different tectonic analyses, e.g. exhumation of lower crustal materials in convergent zones (Cowan and Silling, 1978, Cloos, 1982, 1984). Using this model we explain the deformation path leading to instantaneous shortening along the XY plane of finite strain, resulting in crenulation of early formed foliation in the course of a single, horizontal tectonic movement (Fig. 3.1). We also present numerical models illustrating the spatial distribution of finite strain and shearing movement.

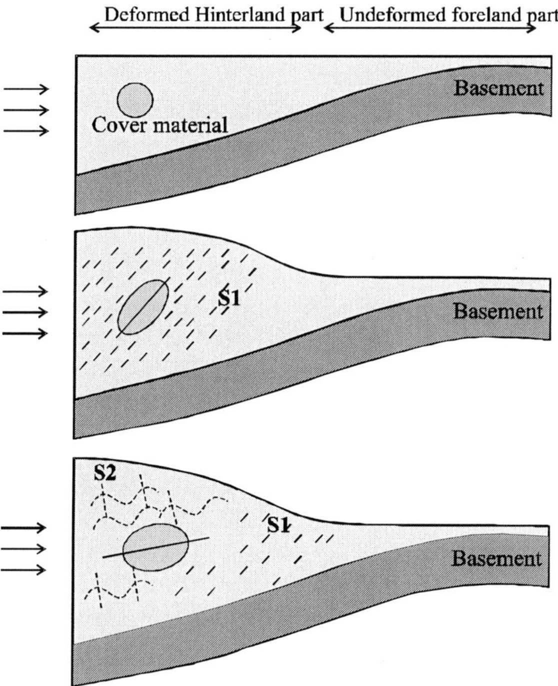

Fig. 3.1 : Schematic presentation of the temporal variation in the local strain field with progressive movement (left to right) in a fold belt. Note progressive development of tectonic fabric (S1) parallel to the XY plane of strain ellipse, and its subsequent folding in response to instantaneous shortening along the XY plane and formation of a new set of fabric (S2).

3.2 Field Observations

3.2.1 Cuddapah Orogenic Belt

The Proterozoic Cuddapah basin of the southern Indian peninsula contains a sedimentary prism, the basin is crescentic in shape and convex towards west, and its eastern margin is marked by a prominent westerly-verging arcuate thrust that seperates the sedimentary cover from the Archean basement (Fig. 3.2a; Naqvi and Rogers, 1987). The Cuddapah strata appear to have suffered regional deformations, as reflected in the development of a variety of structural features, such as folds, cleavage etc. It has also been noted that there is a remarkable variation in structural geometry and strain across the regional trend of the basin (Narayanswamy, 1966). In addition, the Cuddapah strata also record temporal strain variations indicated by overprinting of successive generations of penetrative cleavages as discussed later.

The overall structural disposition of the Cuddapah (*cf.* Meijrink *et al.*, 1984, Kalia and Tewari, 1985, Nagaraja Rao *et al.*, 1987, Matin and Guha, 1996, Mukherjee, 2001, Singh and Mishra, 2002) is comparable to those observed in orogenic wedges that develop dominantly by ductile deformation in response to horizontal shortening from one side (Willett, 1992, Chattopadhyay and Mandal, 2002).

A generalized cross section across the regional trend of the basin is presented in Figure 3.2b (Narayanswamy, 1966). The rocks in the western margin do not have any signature of tectonic deformations. They only show some late joints in the quartzite units. The primary bedding (So) in the rocks dips gently towards east (Fig. 3.3a). Moving progressively towards east, a set of cleavage was observed, which dips easterly at moderate to high angles. The cleavage is axial planar to folds (F1) on primary bedding or layering (Fig. 3.3b,c). The overall structural setting in the Cuddapah orogenic belt indicates that the belt has experienced an overall shortening in the E-W direction. But the direction of principal shortening in local deformations in the western part of the belt is not horizontal, as revealed from the inclined orientations of the penetrative continuous cleavage (S1). The dip of S1 foliation varies towards east. It is more than 60^0 in the western margin, which progressively decrease to about 20^0 close to the eastern margin of the belt (Fig. 3.3d). The nature of the foliation also varies spatially. In the western part it generally occurs as a slaty cleavage, and is progressively replaced by a schistose foliation in the west to east traverse. In addition, the angle between So and S1 also decreases. F1

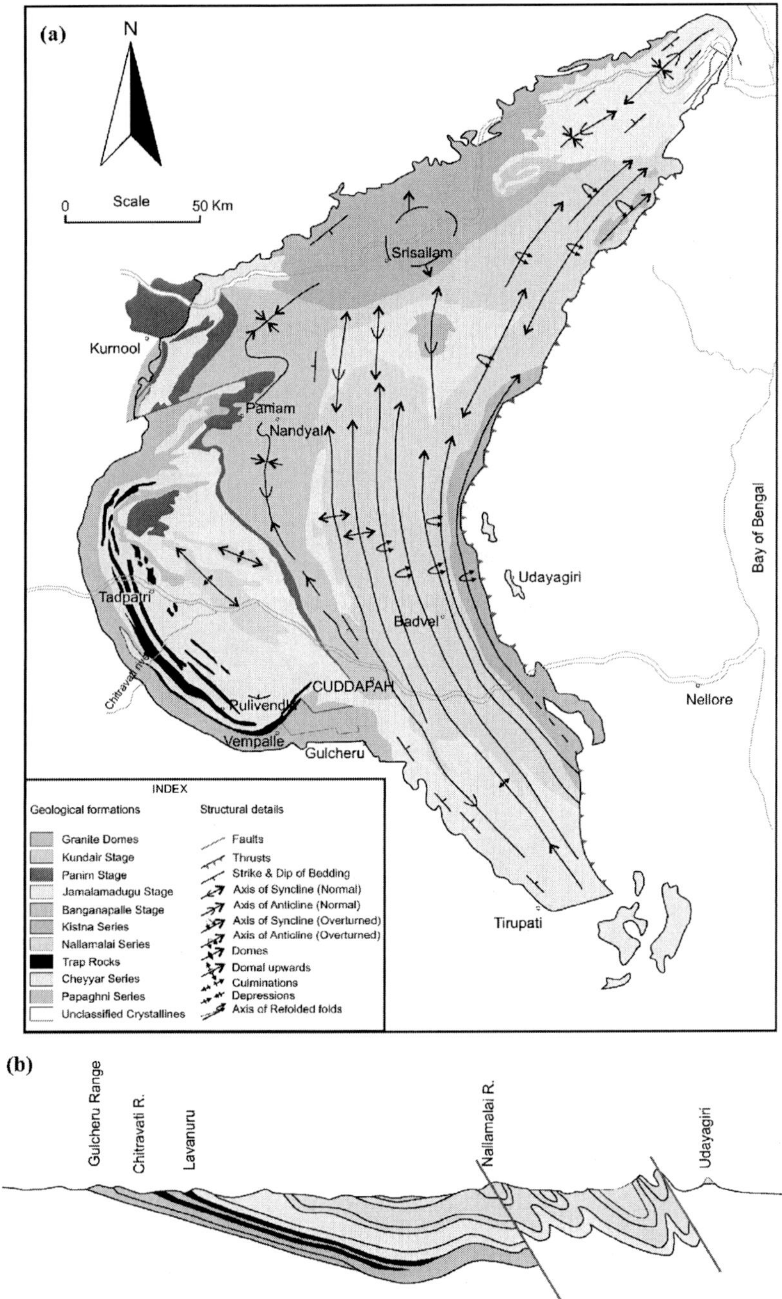

Fig. 3.2a : A geological map of the Cuddapah basin showing the axial traces of map-scale folds (redrawn after Narayanswamy, 1966). **b.** A structural profile across the regional tectonic trend of the Cuddapah fold belt. Note that the tightness of folds decreases in the foreland direction, suggesting decreasing intensity of deformation.

folds in schistose rocks are tight or isoclinal. These are often refolded by later sub-horizontal folds (F2) (Figs. 3.4a and b). F1 and F2 are nearly coaxial. The S1 foliation has been crenulated by F2 folds with axial planes dipping steeply towards east (Fig. 3.4c). A second set of cleavage occurs parallel to the F2 axial planes, which in contrast to S1 is a spaced cleavage. S2 foliation is pervasive all along the eastern margin of the Cuddapah orogenic belt (Fig. 3.4d). S2 has progressively obliterated the first-generation continuous cleavage. In the eastern margin the foliation is marked by segregation of quartz-rich and mica-rich domains, giving rise to a gneissic character in the rocks.

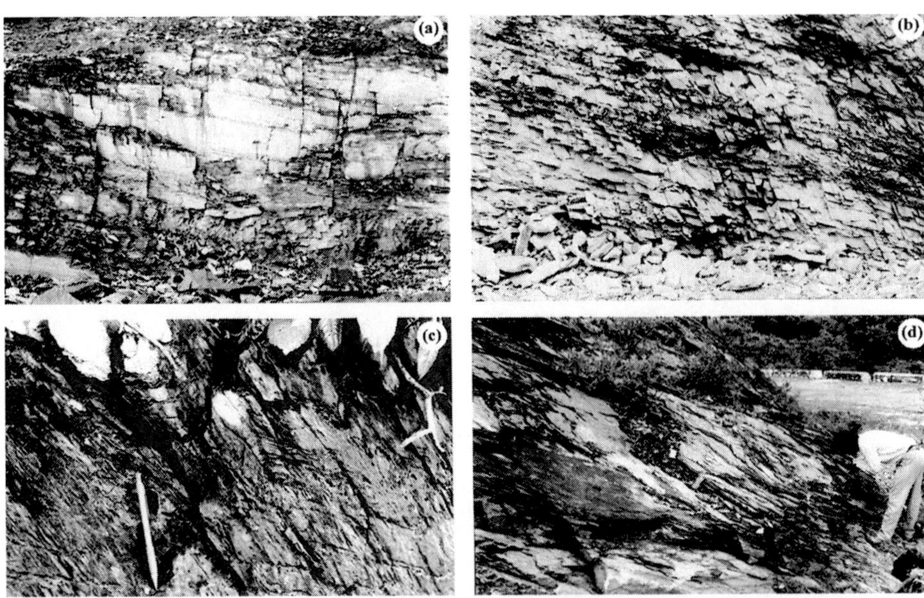

Fig. 3.3 : Variations in the structural style on a west to east traverse in the Cuddapah belt. (a) Easterly dipping beds in the western margin; (b) slaty cleavage in Cumbum Shale, steeply dipping towards west, Mangampeta mines. (c) Westerly-verging first-generation folds (F1) and associated axial plane schistosity (S1), 10 Km from Chitvel on the road to Rapur. (d) Gently dipping S1 schistosity in rocks of the Nallamalli Range exposed on the road to Rapur.

In summary, the strata in the eastern part show westerly verging, tight or isoclinal folds that become open upright folds at the western margin (Narayanswamy, 1966). This structural variation has been reported earlier from a number of orogenic belts (e.g. Nickelsen, 1988, Gray and Mitra, 1993, Boyer and Mitra, 1988). Secondly, this transition from west to east is also characterized by the multiplicity of deformation in the rocks.

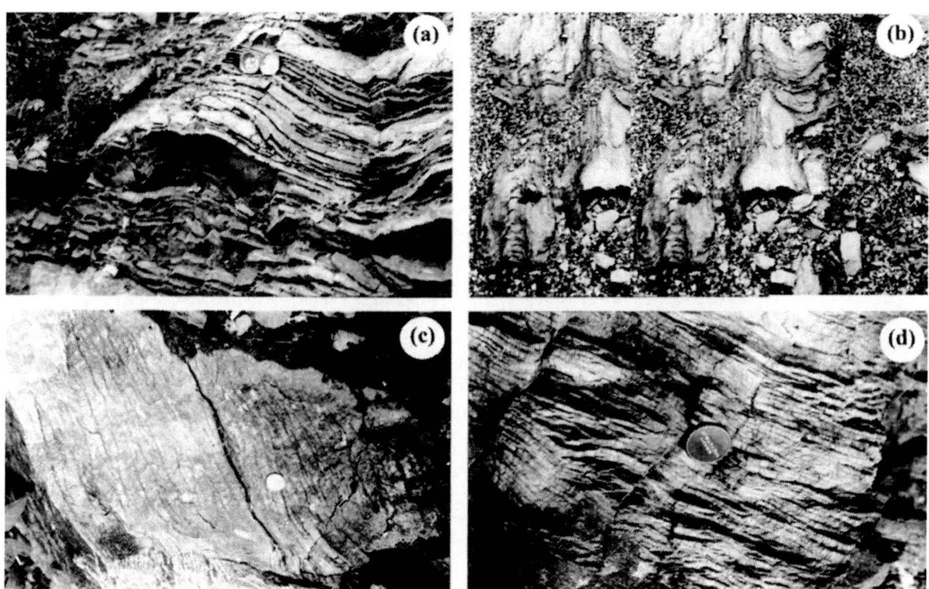

Fig. 3.4 : Structural susperpositions in rocks exposed along the eastern margin of the Cuddapah Mobile belt. (a) Superpostions of late (F2) upright folds on early (F1) recumbent folds. (b) Tight and horizontal upright folds on S1 schistosity. (c) Overprinting of a later set (S2) planar fabric in the form of spaced foliation. (d) Development of crenulations and formation of gneissic character of S2. These structural transitions were observed in the Nallamalai Range on Chitvel to Rapur road sections.

3.2.2 Temporal Variation of Strain

The Cuddapah orogenic belt trends in N-S direction, where the tectonic transport has taken place from east to west. The deformation in the belt is strongly heterogeneous, as revealed from the description in section 3.2. Overprinting of structural fabrics clearly suggests temporal variations in the strain field in the hinterland parts (Fig. 3.5). The strain variations in the tectonic belt have taken place in the same fashion, where the rocks underwent deformation with increasing finite strain, which at one stage of tectonic movement decreased, implying shortening along the XY plane of finite strain. In the following section we present a tectonic model to explain the spatial and temporal variations in local strain under a constant global kinematic framework.

3.3 Tectonic Model

3.3.1 Corner Flow Theory

The ductile crustal deformations in the aforesaid orogenic belt are modeled by considering a viscous slab confined within two converging rigid plates (Fig. 3.6). One of the plates is considered to be dipping, hereafter called plate P_1 and the other is vertical, hereafter called P_2. The viscous material is assumed to be perfectly welded with the rigid plates, and experiences flow as plate P_1 moves against plate P_2, as is the case in the subduction zones located along convergent tectonic boundaries. It may be noted that our consideration of the model base is mechanically different from the frictional detachment surface below thrust sequences localized in the relatively shallow crust. We also assume that there is no volume loss or material accretion in the system during the deformation. The flow in the model is characterized by using Corner flow theory (Batchelor, 1967). The mathematical formulation of the theory is presented below.

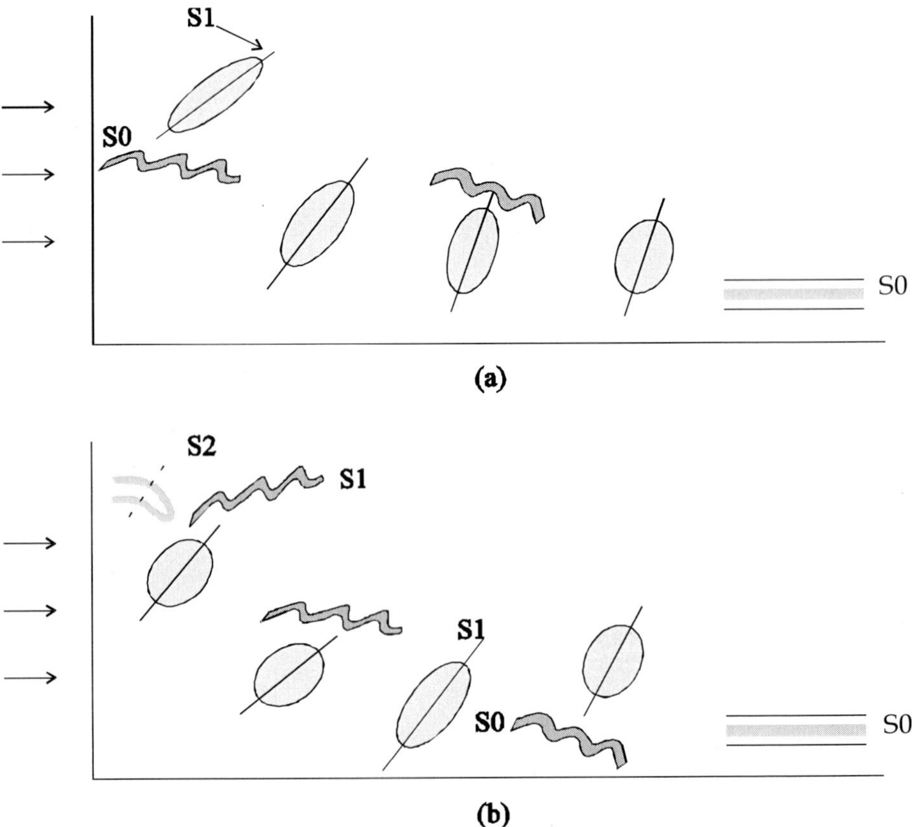

Fig. 3.5a : A generalized scheme of structural styles in relation to local finite strains. **b.** Development of instantaneous shortening along the XY plane of finite strain, resulting in superposition of new folds on the first generation schistosity, as noticed in the Cuddapah.

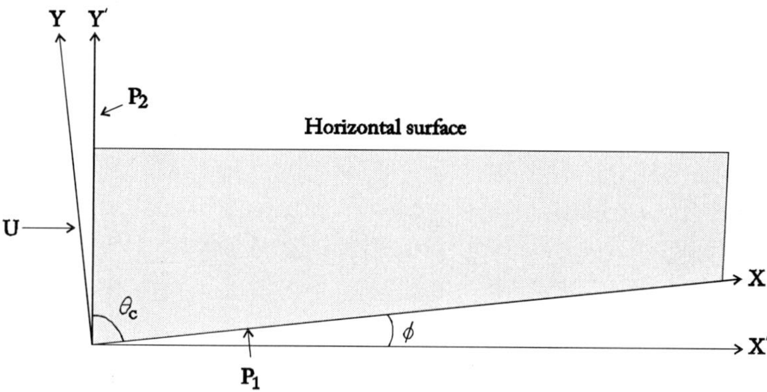

Fig. 3.6 : Considerations of two Cartesian reference frames for describing the flow at the corner between two rigid plates (P1 and P2). P1 approaches P2 with a relative velocity U.

Consider a Cartesian frame *Oxy* with the origin at the intersection point of P_1 and P_2, and *x* axis parallel to plate P_1, dipping at an angle Φ (Fig. 3.6). θ_c is the angle of convergence between the two plates. In the present case $\theta_c = \pi/2 - \Phi$. Plate P_2 moves in the horizontal direction at a velocity *U*, and thus the velocity of plate P_1 relative to plate P_2 is $-U$. We choose an external reference $Ox'y'$ with the x'-axis in the horizontal direction. Following Mandal *et al.* (2002), the instantaneous flow field within the viscous block can be described by velocity components with respect to the polar co-ordinates (r, θ) in the *Oxy* frame as:

$$u_r = A\cos\theta - B\sin\theta + C(\sin\theta + \theta\cos\theta) + D(\cos\theta - \theta\sin\theta) \quad \text{1a}$$

$$u_\theta = -(A\sin\theta + B\cos\theta + C\theta\sin\theta + D\theta\cos\theta) \quad \text{1b}$$

where,

$$A = \frac{U}{\sin^2\theta_c}\left[\cos^2\theta_c\cos\phi - K(\theta_c + \sin\theta_c\cos\theta_c)\right] \quad \text{2a}$$

$$B = U\sin\phi \quad \text{2b}$$

$$C = UK \quad \text{2c}$$

$$D = -\left[\cos\phi + \frac{1}{\sin^2\theta_c}\left\{\cos^2\theta_c\cos\phi - K(\theta_c + \sin\theta_c\cos\theta_c)\right\}\right]U \quad \text{2d}$$

$$K = \frac{\theta_c - \sin\theta_c \cos\theta_c \left\{(1-\theta_c)^2 \cos\phi + \theta_c^2\right\} + \sin^2\theta_c \left\{\theta_c\left(\cos\phi - 1\right) + \sin\phi\right\}}{\theta_c^2 - \sin^2\theta_c} \quad 2e$$

With the help of velocity functions Eqs. (1a) and (1b) it is possible to simulate the flow in the viscous medium for incremental relative displacements between two boundary plates. Considering the following coordinate transformations:

$$\begin{bmatrix} x \\ y \end{bmatrix} = \begin{bmatrix} \cos\phi & \sin\phi \\ -\sin\phi & \cos\phi \end{bmatrix} \begin{bmatrix} x' \\ y' \end{bmatrix} \quad 3$$

$$\begin{bmatrix} u' \\ v' \end{bmatrix} = \begin{bmatrix} \cos(\phi+\theta) & -\sin(\phi+\theta) \\ \sin(\phi+\theta) & \cos(\phi+\theta) \end{bmatrix} \begin{bmatrix} u_r \\ u_\theta \end{bmatrix} \quad 4$$

With the help of the corner flow theory we analyzed in the following way the strain distribution pattern and its temporal variation in convergent tectonic belts. A rectangular block was chosen with a horizontal base resting upon a rigid base ($\Phi = 2^o$), and buttressed by a rigid vertical wall from one side (i.e. $\theta c = 90^o$). Small circles (passive marker) were drawn on the block with their centers arranged in a Cartesian grid (Fig. 3.7). The horizontal basal plate was moved against the vertical plate at a constant velocity to set a flow in the viscous block. Considering incremental displacements of points on circular passive markers (derived by using eqs. 3 and 4), the deformed shapes of circles were determined for increasing finite displacements of the moving basal plate. Deformed circles reveal a heterogeneous strain distribution with finite strain varying both in the horizontal and vertical directions. The finite strain decreases continuously from the hinterland to foreland, as noticed in the Cuddapah orogenic belt. The XY planes of strain ellipses over most of the region verge in the foreland direction. This indicates that the tectonic fabric (foliation parallel to XY plane of finite strain) develops essentially dipping in the hinterland direction even the orogenic belts evolve through a regional horizontal contraction. The model results explain the development of hinterland-dipping first-generation schistosity in the frontal parts of the Cuddapah (Figs. 3.7b-g). Fyson (1971) suggested that the fold style in an orogenic belt changes from upright at shallow depth to recumbent at deeper level. This indicates that the XY planes should be nearly vertical close to the surface and should gradually become horizontal at depth.

However, our models suggest that the depth versus fold-style relation may not be valid everywhere in the orogenic belt. In the hinterland part the XY planes are sub-horizontal at shallow level whereas steeply-dipping at deeper level.

Using the corner flow theory we also investigated another important issue concerning superposed deformations observed in the orogenic belt described in section 3.2 (Fig. 3.7). It has been seen that a tectonic fabric which has developed parallel to XY plane of finite strain undergoes folding during subsequent stages of deformation. The superposition of folding process indicates a temporal variation of the strain field. Employing this theoretical model we inspect the nature of temporal variation in finite strain at different locations of a convergent belt (Fig. 3.8). For this analysis, *deformation paths* at chosen locations were determined taking into account the finite strains at different time instants. The paths in the hinterland of an orogenic belt are unsteady, which we will consider here. In the deeper level the finite strains at points located closer to the hinterland boundary increases continuously (Fig. 3.8b). Moving in the shallower direction, we obtain a different deformation path, where the finite strain increases, but unsteadily (Figs. 3.8c and d). There is a slight drop in strain, implying an instantaneous shortening along the XY plane of finite strain developed at the earlier stages of convergence movement. The path explains that any tectonic fabric formed parallel to the XY plane in the deeper region can undergo shortening, and develop folds in the course of a single, continuous contraction, as noticed in the Cuddapah orogenic belt. All the paths in the middle level keep positive gradients, implying continuous increase in finite strain. The analysis suggests that a tectonic fabric formed parallel to the XY plane is not likely to fold at this depth (about 20 to 30 Km). The deformation paths at the shallow depths show contrasting variations in finite strain. In the central region of hinterland the paths strongly plunge with negative gradients, whereas those on either side of this location maintain positive gradients. The negative gradient of deformation paths suggest that the XY plane must involve instantaneous shortening, resulting in decrease of finite strain. It thus appears from the strain analysis that superposed deformations in a orogenic belt can occur in the course of a single, continuous convergence movement.

During the evolution of an orogen shearing movement develops, which re-orient earlier structures, such as folds, foliations, observed in physical experiments (Chattopadhaya and Mandal, 2002). In this study we investigated the distribution of vertical and horizontal shear in a orogenic belt (Fig. 3.9). Models containing two sets of horizontal and vertical lines were deformed by using the velocity functions in Eqs. 3 and 4. The distortion patterns of original perpendicular lines reveal spatial variations of finite shear. In the frontal part top-to-foreland shearing movement is dominant, resulting in deflection of vertical lines in the foreland direction. This explains localization of foreland-vergent tectonic fabrics and folds in the Cuddapah orogenic belt. On the other hand, the shear is strongly heterogeneous in the hinterland, and acts dominantly in the vertical direction. Planar structural elements, such as foliation, bedding, are thus likely to steepen its dip in response to the vertical shear (Ramsay and Huber, 1987). There is a zone of no-finite shear that separates the two zones of vertical and horizontal shear (Fig. 3.10).

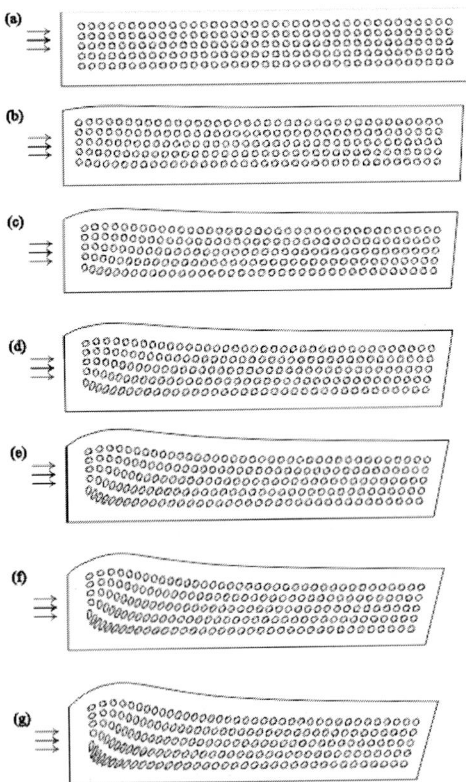

Fig. 3.7 : Successive stage of numerical models, based on the Corner flow theory, showing the heterogeneous strain field and its temporal variation in a mobile belt. Model contraction was from left to right. (Here the basal inclination i.e. ϕ is taken 2° in computation).

The corner flow models employed in this study do not take into account the effect of gravity in the orogenic belts. However, the effect becomes significant with progressive crustal thickening, which we have investigated using finite element models presented in the following section.

3.3.2 Finite Element Approach

The finite element models were developed using the ANSYS software code. The initial model had wedge geometry, which was simulated considering the displacements (12 % overall shortening) from equations 1 (a) and (b). All the co-ordinates following displacements were obtained by using the velocity functions (Eqs. 3 and 4), and the displaced positions were then used for developing the finite element code employing MATLAB, and the FE simulations were carried out to derive further positional changes in presence of the gravity, and to determine the strain distributions. In finite element modeling we consider the materials constituting the orogenic belt as Maxwell viscoelastic with realistic density (2750 kg/cubic m), Poisson's ratio (0.25), shear modulus (10^{11} Pa) and viscosity (10^{21} Pa s). The entire system is subjected to a gravity field with g = 9.8 ms^{-2}. Further horizontal shortening is simulated through transient dynamic finite element code, taking realistic shortening time (in the order of Myr). In this case the model undergoes deformations under the influence of both horizontal stresses and gravity forces. We designed the FE model with fine meshing, consisting of 57501 nodes in order to obtain higher resolution in the strain field (Fig. 3.11).

With the finite element method we studied the distribution of finite stain in an orogenic belt (Fig. 3.12). The overall strain pattern does not differ much from those obtained from the corner flow theory discussed earlier. However, the XY planes of strain ellipse in the hinterland of the wedge were mostly horizontal, indicating the flattening effect of gravity. We also estimated the spatial variations in the principal strain and finite shear developed during the phase of movement involving gravity (Fig. 3.13). Models results show that the magnitudes of principal strains are highest in the shallower levels, and decreases with increasing depth (Fig. 3.13a). On the other hand, models show localization of maximum shear along an inclined zone in front of the elevated part of wedges (Fig. 3.13b).

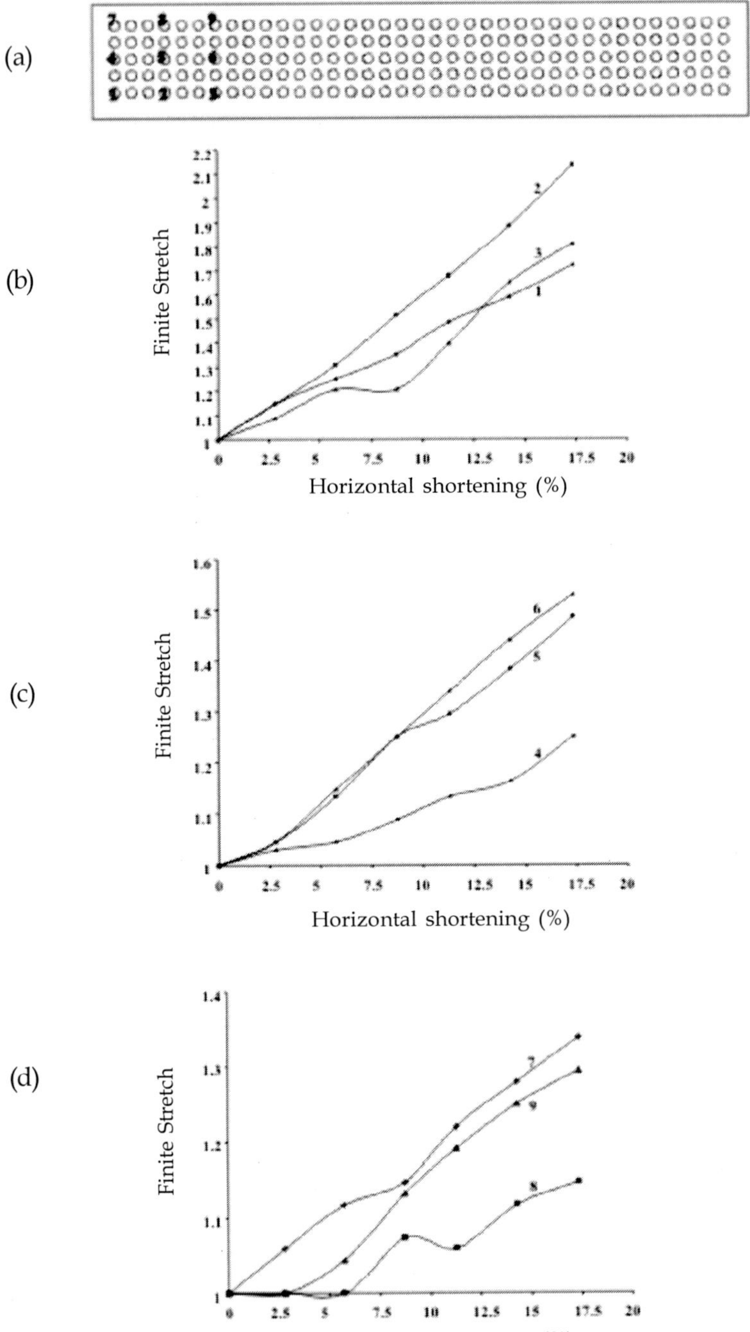

Fig. 3.8 : Variations in finite strain at different locations of a mobile belt as a function of overall contraction in percentage. See text for details.

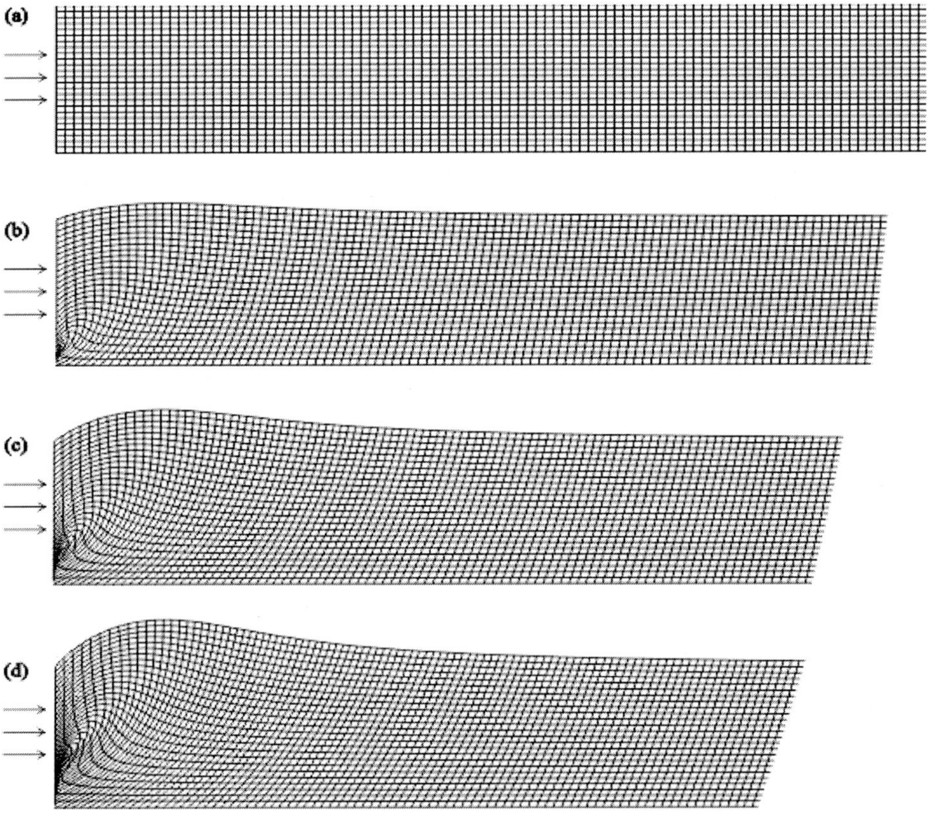

(a)

(b)

(c)

(d)

Fig. 3.9 : Distortions patterns of Cartesian grids in numerical models showing the distribution of horizontal and vertical finite shear in successive stages of progressive horizontal contraction.

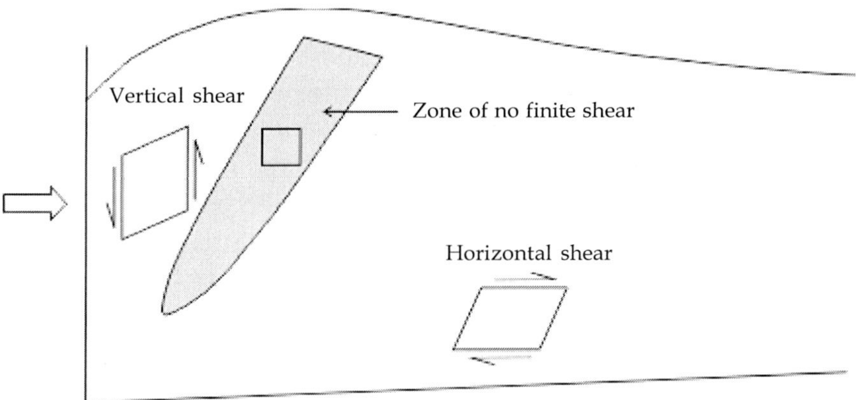

Fig. 3.10 : A schematic sketch illustrating the fields of horizontal and vertical shear with opposite senses separated with the zone of no finite shear.

3.4 Discussion and Conclusions

3.4.1 Interpretation of Regional Ductile Deformations in Orogenic Belts

Physiographic features and structural styles in the Cuddapah orogenic belt indicate that the crustal segment has experienced an overall horizontal contraction from one side. The intensity of deformation decreases from the hinterland to foreland direction and the crustal materials lying in front of the orogenic belts have remained virtually undeformed. The corner flow model presented in this study reveals that in such a dynamic setting the horizontal contraction must be associated with top-to-foreland shear in horizontal direction (*cf.* Hafner, 1951), as reflected from foreland-vergent, first-generation penetrative S1 foliation in the two orogenic belts. On the other hand, numerical simulations indicate that intense vertical shear is likely to develop in the hinterland area during the horizontal contraction movement. Thus, local strains in both the foreland and hinterland domains are likely to be rotational in nature but with opposite senses. Our study suggests that an orogenic belt can evolve through a complex strain history even the broad tectonic framework is simple, involving a single, continuous horizontal contraction.

In the eastern margin of the Cuddapah belt the rocks are multiply deformed, showing overprinting of a spaced, crenulation cleavage over the earlier continuous foliation. According to a homogeneous deformation model, a foliation developing parallel to the XY plane of finite strain is unlikely to be shortened in the same movement (here it is horizontal contraction), as it remains in the extensional field of deformation. It is thus evident that the conditions of deformation at any location must have changed temporally, giving rise to shortening along the XY plane at a later stage of movement. Our study suggests that such a kinematic change is possible to explain employing corner flow tectonic model. The strain paths obtained from the model show that the finite strain at a point located in the hinterland increases to a maximum, but decrease afterward, implying that there occurs incremental shortening along the XY-plane of finite strain in the course of deformation, leading to a new set of foliation on the earlier one (Fig. 3.5). Successive structural developments in orogenic belts thus may not be the results of discrete deformation events, but might have taken place in the course of a single, continuous tectonic movement.

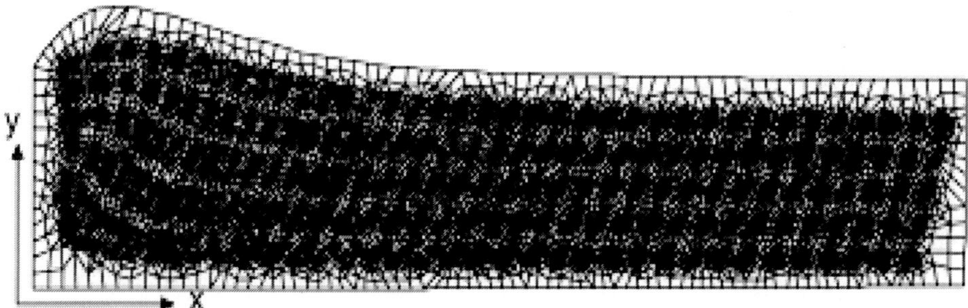

Fig. 3.11 : Initial model designed for finite element analysis. The meshes are extremely fine in the central part of the model. The initial configuration of the model was obtained from the Corner flow theory, considering an overall shortening by 12%.

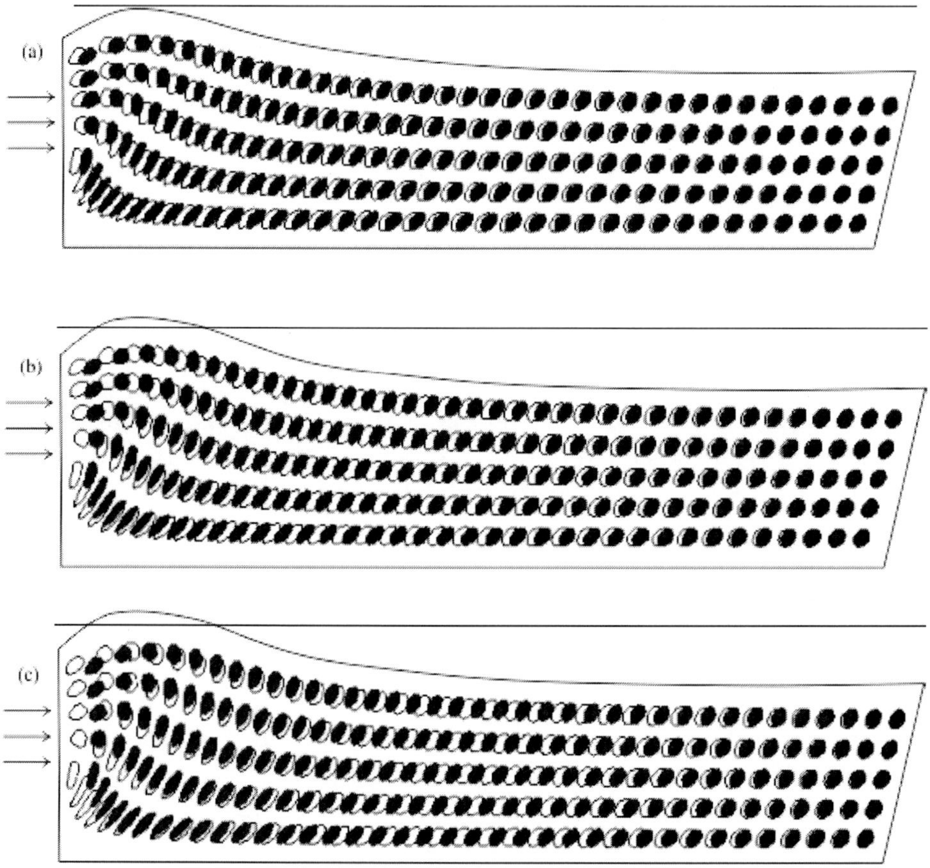

Fig. 3.12 : Strain distributions in finite element models. Open ellipses indicate finite strains obtained from the Corner flow theory, and solid ellipses show their changes during horizontal movement in the presence of gravity.

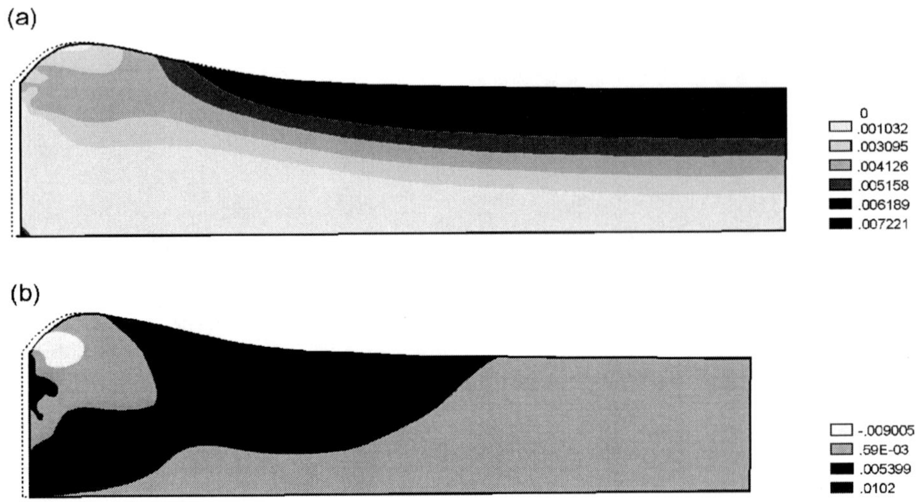

Fig. 3.13 : Spatial variations in the magnitudes of (a) principal finite strain (λ_1) and (b) shear in finite element models (γ).

In this study we have dealt with two stages of deformation in the orogenic belts. Numerical models indicate that the timing of sequential deformation episodes varies spatially. For example, there occurs decreasing finite strain in the hinterland, implying superposition of shortening along the XY plane of finite strain (i.e. development of second-generation folds on the earlier fabric), whereas the finite strain develops in the frontal part, suggesting formation of the first phase of deformational structures. It then appears that time of the second phase of deformation in one locality can be the time of first phase of deformation elsewhere. Regional field studies reveal structural sequences produced under more than two deformation episodes. Time correlation of such multiple deformations observed in different locations in an orogenic belt is thus likely to be more complex. The model presented here does not give any account of such complex deformation history. Intuitively, we propose that interactions of other geological processes, e.g. emplacement of intrusive bodies, formation of thrusts etc., can result in additional effects on the strain field, and lead to increasing multiplicity of deformation episodes.

3.4.2 Structural Analysis of Active Orogenic Belts

We have shown from a Precambrian orogenic belt that the ductile structures in asymmetric collisional tectonic zones develop with a systematic spatial variation of deformation. The same deformation style can occur in Phanerozoic orogenic belts. To test this, we can compare our findings with the structural sequence in the Darjeeling-Sikkim Himalaya. The orogenic belt contains virtually undeformed and unmetamorphosed rocks in the mountain front on south, and shows increasing deformation and metamorphism in the north direction (Acharyya, 1975, Acharyya and Ray, 1977). The deformed rock sequence is intervened by a number of thrusts (Acharyya, 1975, 1994, 1999). In this discussion we will, however, focus mainly upon penetrative ductile deformations that prevailed globally and varied continuously in the N-S transect. This context excludes local structural complexities resulting from localization of discontinuous deformations in response to relatively small-scale responses.

Siwalik rocks that constitute the mountain front are apparently undeformed, and show gently-dipping primary beds. The deformation becomes more and more pronounced towards north. The Gondwana rocks lying north of the Siwalik unit are macroscopically deformed, developing south-vergent schistosity in the pelitic rocks. The primary belds are strongly folded (Acharyya, 1975, Acharyya and Roy, 1977). This is followed by the Daling rocks in the north, which also show a dominant south-vergent penetrative schistosity. Superposition of a later tectonic fabric, and associated folds were noticed in the Daling rocks exposed further north of Kalijhora. This structural superposition could be observed over the entire section from Kalijhora to Tista Bazar. The structural transitions on the N-S transects resemble that observed in the Cuddapah mobile belt. We thus infer that the deformation pattern derived from the Corner flow theory can also be used to explain the structural styles in Phanerozoic orogenic belts.

3.4.3 Modelling Approach

In this section we will discuss briefly to address why a viscous model has been adopted to analyze the spatial and temporal variations in the Cuddapah Orogenic belt. Different tectonic models that are available in the literature can be grouped broadly into three rheological categories: 1) *Coulomb model*, 2) *Elastic-plastic model* and 3) *Viscous models*. The application of a rheological model must be consistent with the actual rheology of crustal materials, which is again a function of both time and space scales. This section presents an overview on this issue.

Coulomb rheology is governed mainly by intergranular frictional dynamics. Thus, the formation of deformational structures in materials with Coulomb rheology depends on the stress, but not the strain rate. Field evidences suggest that only the upper thin part of the crust behave approximately as Coulomb layer (McClay and Ellis, 1987). Coulomb rheological models are therefore appropriate to analyze brittle deformations taking place in the shallow crust. In tectonic studies many workers have used the Coulomb rheology to investigate the evolution of thrust systems that are common in the frontal mountain belts (Davis *et al.*, 1983, Dahlen, 1990). It may be noted that we cannot analyze penetrative ductile deformations in rocks considering a Coulomb model. On the other hand, elastic-plastic models appear to be more appropriate to take into account the effect of ductile strains in rocks. An elastic substance yields at a critical stress, giving rise to plastic strain in the form of permanent ductile strain. However, physical and numerical experiments show that plastic strains develop always heterogeneously in the body, forming discrete plastic (high ductile strain) zones (Bowden and Raha, 1970, Anand and Spitzig, 1980, Tvergaard *et al.*, 1981, Stockmal, 1983, Willett, 1999, Mandal *et al.*, 2004), which are often observed in tectonic belts. But deformation structures in the orogenic rocks show ductile strains being distributed over the entire orogenic belt, as observed in the Cuddapah belt, which is difficult to explain considering an elastic-plastic rheology. Geological and geophysical evidence indeed suggest that the crust, in overall, behave as a visco-elastic layer on a long time scale, and they flow approximately with viscous rheology at slow rates of deformation (Ranalli, 1987, Turcotte and Schubert, 2001). The main objective of our study is to show the pattern of continuous variation in (long-time scale) ductile strain across an orogenic belt, excluding local effects resulting from thrusts (Coulomb deformation) or localized strain (elastic-plastic deformation). This makes the basis for choosing a viscous rheological model.

3.4.4 Limitations of the Kinematic Model

In this modeling we approach with the thick-skined tectonics of crustal deformation in orogens. Our model thus explains the spatial and temporal variations of structures produced by continuous viscous flow of the crustal materials in convergent tectonic regimes. However, the uppermost thin (6-8 km) part of the crust can behave differently, and shows predominance of deformation localization, forming faults and shear zones, as discussed in the previous section. The geometrical dispositions of ductile structures can be re-oriented and modified locally by the process of localized deformation, like thrusting. The viscous model used in this study evidently does not take into account such local structural variations in orogens.

There are several other limitations in this modeling approach. First, the material is assumed to be rheologically homogeneous (averaged) and spatially continuous, and thereby undergoes continuous deformation in response to the movement in the basal rigid plate. In contrary, geophysical and petrological evidences suggest that the physical and chemical conditions in orogens vary in both the horizontal and vertical directions. As a consequence, the crustal materials in different parts are likely to be rhelogically different. In many orogenic belts deformational processes can be complex due to strong rheological variations (Willett, 1999, Beaumont *et al.,* 2001, Groom *et al.,* 2008). The homogeneous viscous model presented in this study needs to be advanced to deal with that kind of rheologically heterogenous orogenic systems. Secondly, all the numerical simulations were run at a constant angle of the convergent plates. However, the convergent angle can change and thereby influence the flow pattern of material in the convergent zone. Thirdly, the Corner flow model does not take into account the effect of gravity. Finite element models, however, indicate that the gravity has little effects on the strain field obtained from the corner flow theory. The boundary conditions employed in the corner flow model are somewhat idealistic. The viscous wedge has been assumed to be coherent to the subducting and the overriding solid plates, allowing no slip during the tectonic movement. However, there can be slip at the plate boundaries due to several geological conditions, e.g. localization of shear failure. In such situations the strain distribution can be somewhat different from that described here.

3.5 Conclusions

Based on our field observations and theoretical models, we come to the following conclusions. 1) Deformational structures in orogenic belts develop in response to heterogeneous strain fields even when the belts evolve through simple horizontal contraction. The strain fields vary temporally, and result in multiple deformation episodes in the course of a single, continuous contraction movement. 2) In the hinterland of orogenic belts there occurs instantaneous shortening locally along the XY plane of finite strain, as reflected from folding on the first generation fabric in the Cuddapah orogenic belt and the Himalayan mountain blet. 3) Orogenic belts undergoing contraction from one side will be essentially associated with top-to-foreland shearing on horizontal plane and intense shearing on vertical planes in the hinterland. The two regions are separated by an inclined zone with no finite shear. 4) The role of gravity does not influence the strain field significantly during the period of active horizontal contraction.

Acknowledgements

We thank Professors Gautam Mitra and Dilip Mukhopadhaya for their insightful and incisive comments on the manuscript. The work has been supported by the Department of Science and Technology, India, sanctioned to NM. AKM and MM are grateful to the CSIR and ICCR respectively for proving them with a research fellowship.

Bibliography

Acharyya, S.K. (1975). On the nature of the main Boundary Fault in the Darjeeling Sub-Himalaya, *Geological Survey of India, Misc.Pub. no.* 24(2), pp. 395-406.

Acharyya, S.K. and Ray, K.K. (1977). Geology of the Darjeeling-Sikkim Himalaya, Guide to excursion no.4.Fourth International Gondwana Symposium, Calcutta, India, *Geological Survey of India*, pp. 1-25.

Acharyya, S.K. (1994). The Cenozoic foreland basin and tectonics of the Eastern Sub-Himalaya: Problems and prospects, *In: Siwalik foreland basin of Himalaya, Kumar, R., Ghosh, S. and Phadtare, N.R. (eds), Himalayan Geology*, 15, pp. 3-21.

Acharyya, S.K. (1999). Tectonic evolution of the eastern Himalayan Tertiary basin, *In: Geodynamics of NW Himalaya., Jain, A.K. and Manickavasagam, R.M. (eds), Gondowana Research. Memoir*, 6, pp. 263-271.

Anand, L. and Spitzig, W.A. (1980). Initiation of localised shear bands in plane strain, *Journal of Mechanics and Physics of Solids*, 28, pp.113–128.

Batchelor, G. K. (1967). An Introduction to fluid dynamics, *The University Press, Cambridge*. pp. 615.

Beaumont, C., Jamieson, R. A., Nguyen, M. H. and Lee, B. (2001). Himalayan tectonics explained by extrusion of a low-viscosity crustal channel coupled to focused surface denudation, *Nature*, 414, pp. 738 – 742.

Bowden, P.B. and Raha, S. (1970). The formation of micro-shear bands in polystyrene and polymethylmethaacrylate, *Philosophical Magazine*, 22, pp. 463–482.

Boyer, S. E. and Mitra, G. (1988). Relationship between deformation of crystalline basement and sedimentary cover at the basement/cover transition zone of the Appalachian Blue Ridge Province, *In: Mitra, G. and Wojtal, S., eds, Geometry and Mechanisms of Thrusting, with special reference to the Appalachians, Geological Society, America. Special. Paper*, 222, pp. 119-136.

Chapple, W.M. (1978). Mechanics of thin-skinned fold-and-thrust belts, *Geological Society of America, Bullatin*, 89, pp. 1189–1198.

Chattopadhyay, A. and Mandal, N. (2002). Progressive changes in strain patterns and fold styles in a deforming ductile orogenic wedge: an experimental study, *Journal of Geodynamics*, 33, pp. 353-376.

Cloos, M. (1982). Flow melanges: Numerical modeling and geologic constrants on their origin in the Franciscan subduction complex, California, *Geological Society of America, Bullatin*, ,93, pp. 330-345.

Cloos, M. (1984). Flow melanges and the structural evolution of accretionary wedges, *Geological Society of America, Special paper*, 198, pp. 71-79.

Cowan, D. S. and Silling, R. M. (1978). A dynamic, scaled model of accretion at trenches and its implications for the tcctonic evolution of subduction complexes, *Journal of Geophysical Research*, 83, pp. 5389-5396.

Dahlen, F.A. (1984). Noncohesive critical Coulomb wedges: An exact solution, *Journal of Geophysical Research*, 89, pp. 10125–10133.

Dahlen, F.A. (1990). Critical taper model of fold-and-thrust belts and accretionary wedges, *Annual Rev. Earth Planetary Science*, 18, pp. 55–99.

Davis, D., Suppe, J. and Dahlen, F.A. (1983). Mechanics of fold-and-thrust belts and accretionary wedges, *Journal of Geophysical Research*, 88, pp. 1153–1172.

England, P. and G.Houseman (1985). Role of lithospheric strength heterogeneities in the tectonics of Tibet and neighboring regions, *Nature*, 315, pp. 297 - 301.

Fyson, W. K. (1971). Fold attitudes in metamorphic rocks, *American Journal of Science*, 270, pp. 373-382.

Gray, M. B. and Mitra, G. (1993). Migration of deformation fronts during progressive deformation: Evidence from detailed structural studies in the Pennsylvania Anthracite region, *Journal of Structural Geology*, 15, pp. 435-449.

Grujic, D., Casey, M., Davidson, C., Hollister, L. S., Kündig, R., Pavlis, T. and Schmid, S. (1996). Ductile extrusion of the Higher Himalayan Crystalline in Bhutan: evidence from quartz microfabrics, *Tectonophysics*, 260, pp. 21-43.

Ghosh, S. K. (1982). The problem of shearing along axial plane foliations, *Journal of Structural Geology*, 4, pp. 63-67.

Ghosh, S. K. (1993). Structural Geology: Fundamentals and Modern Developments, *Pergamon Press. England*. pp. 170.

Groome, W. G., Koons, P. O. and Johnson, S. E.(2008). Metamorphism, transient mid-crustal rheology, strain localization and the exhumation of high-grade metamorphic rocks, *Tectonics*, 27, TC1001, doi:10.1029/2006TC001992.

Hafner, W. (1951). Stress distribution and faulting, *Bullatin of Geological Society of America*, 62, pp. 373-398.

Jime´nez-Munt, I. and Platt, J. P. (2006). Influence of mantle dynamics on the topographic evolution of the Tibetan Plateau: Results from numerical modeling, *Tectonics*, 25, TC6002, doi:10.1029/2006TC001963.

Kalia, K. L.and Tewari, H. C. (1985). Structural trends in the Cuddapah Basin from Deep Seismic Soundings (DSS) and their tectonic implication, *Tectonophysics*, 115, pp. 68-86.

Macaya, J., Gonzalez-Lodeiro, F., Martinez-Catalan, J. R. and Alvarez, F. (1991). Continuous deformation, ductile thrusting and backfolding of cover and basement in the Sierra de Guadarrama, Hercynian Orogen of central Spain, *Tectonophys*, 191, pp. 309-321.

Mandal, N., Samanta, S. K. and Chakraborty, C. (2002). Flow and strain patterns at the terminations of tapered shear zones, *Journal of Structural Geology*, 24, pp. 297-309.

Mandal, N., Misra, S. and Samanta, S. K. (2004). Role of weak flaws in nucleation of shear zones: an experimental and theoretical study, *Journal of Structural Geology*, 26, pp. 1391-1400.

Matin, A. and Guha, J. (1996). Structural geometry of the rocks of the southern part of the Nallamalai Fold Belt, Cuddapah Basin, Andhra Pradesh, *Journal of Geological Society of India*, 47, pp. 535-545.

McClay K R and Ellis P G. (1987). Geometries of extensional fault systems developed in model experiments, *Geology*, 15, pp. 341-344.

Meijrink, A. M. J., Rao, D. P. and Rupke, J. (1984). Stratigraphy and structural development of the Precambrian Cuddapah Basin, S. E. India, *Precambrian Research*, 26, pp. 57-104.

Mukherjee, M. K. (2001). Structural pattern and kinematic framework of deformation in the southern Nallamalai fold-fault belt, Cuddapah district, Andhra Pradesh, Southern India, *Journal of Asian Earth Science*, 19, pp. 1-15.

Nagaraja Rao, B. K. Rajurkar, S. K. Ramlingaswami and G. Ravindra Babu. (1987). Stratigraphy, structure and evolution of the Cuddapah basin, *In: Purana Basins of Peninsular India. Geological Society of India, Memoir*, 6, pp. 33-86.

Naha, K. and Mohanty, S. (1988). Response of basement and cover rocks to multiple deformation: a study from the Precambrian of Rajasthan, Western India, *Precambrian Resesearch*, 42, pp. 77-96.

Naqvi, S. M. and Rogers, J. J. W. (1987). Precambrian geology in India, *Oxford University Press, Oxford.*

Narayanswamy, S. (1966). Tectonics of the Cuddapah Basin, *Journal of Geological Society India, 7*, pp. 33-50.

Nickelsen, R. P. (1988). Structural evolution of folded thrusts and duplexes on a first-order anticlinorium in the valley and Ridge Province of Pennsylvania, *In: Mitra, G., Wojtal, S. F.(Eds), Geometries and Mechanism of thrusting, with Special reference to the Appalachians. Special paper-Geological Society of America,* 222, pp. 89-106.

Platt, J.P. (1986). Dynamics of orogenic wedges and the uplift of high-pressure metamorphic rocks, *Geological Society of America, Bullatin*, 79, pp. 1037–1053.

Ramberg, H. (1975). Particle paths, displacements and progressive strain applicable to rocks, *Tectonophysics*, 28, pp. 1-37.

Ramsay, J. G. (1967). Folding and fracturing of rocks, *McGraw-Hill, New York.*

Ramsay, J. G.and Huber, M. I. (1987). The techniques of modern structural geology. 2. Folds and fractures, *Academic Press, London.*

Ranalli, G. (1987). Rheology of the Earth, *Allen and Unwin, Boston.*

Rossetti, F., Faccenna, C., Ranalli, G. and Storti, F. (2000). Convergence rate-dependant growth of experimental viscous orogenic wedges, *Earth and Planetary Science Letters*, 178, pp. 367-372.

Royden, L. (1996). Coupling and decoupling of crust and mantle in convergent orogens: Implications for strain partitioning in the crust, *Journal of Geophysical Research*, 101, pp. 17679–17705.

Sanderson, D. J. (1979). The transition from upright to recumbent folding in the Variscan fold belts of southwest England: a model based on the kinematics of simple shear, *Journal of Structural Geology*, 1, pp.171-180.

Singh, A. P. and Mishra, D. C. (2002). Tectonosedimentary evolution of Cuddapah basin and Eastern Ghats mobile belt (India) as Proterozoic collision: gravity, seismic and geodynamic constrants, *Journal of Geodynamics*, 33, pp. 249-267.

Stockmal, G. S. (1983). Modelling of large-scale accretionary wedge deformation, *Journal of Geophysical Research*, 88, pp. 8271-8287.

Sutton, J. and Watson, J. (1955). The structure and stratigraphic succession of the Moines of Fannich Forest and Strath Bran, Ross-shire, *Quarternary Journal of Geological Society, London*, 110, pp. 21-53.

Sutton, J., Watson, J. (1959). Structures in the Caledonides between Loch Duich and Glenelg, North-West Highlands, *Quarternary, Journal of Geological Society of London*, 114, pp. 231-257.

Turcotte, D.L. and Schubert, G. (2001). Geodynamics, *Wiley, New York.*

Tvergaard, V., Needleman, A. and Lo, K.K. (1981). Flow localization in the plane strain tensile test, *Journal of Mechanics and Physics of Solids*, 29, pp. 115–142.

Twiss, R. J. and Moores, E. M. (2007). Structural Geology (2nd Ed), *W. H. Freeman and Company, New York.*

Wang, W.-H. and Davis, D.M. (1996). Sandbox model simulation of forearc evolution and non-critical wedges, *Journal of Geophysical Research*, 11, pp. 11329–11339.

Weiss, L. E., McIntyre, D. B. and Kursten, M. (1955). Contrasted styles of folding in the rocks of Ord Ban, Mid- Strathspey. *Geology Magazine*, 92, 21-36.

Willett, S. D. (1992). Dynamics and kinematic growth and change of a Coulomb wedge, *In McClay, K. R. (Ed), Thrust Tectonics. Chapman & Hall, London*, pp. 19-31.

Willett, S. D. (1999). Rheological dependence of extension in wedge models of convergent orogens, *Tectonophysics*, 305, pp. 419–435.

CHAPTER 4

Grain Size Distribution Patterns of Some Rivers in the Light of Three Model Distributions

Barendra Purkait

Abstract

Grain size distribution patterns of five rivers – Hooghly, Bhagirathi, Damodar, Ajoy and the Usri, were critically analyzed and compared with three model distributions: log-normal, log-hyperbolic and log-skew-Laplace. The log-normal and log-skew-Laplace are the limiting distributions of log-hyperbolic family. The Hooghly-Bhagirathi river system from where the sand samples were collected is tidal in nature whereas the other three rivers are non-tidal. The tidal effect on grain size distribution patterns was also studied. Bed load samples were collected from the foreset of different bedform sizes. A total number of 126 sediment samples were collected from different sections of the rivers under study.

It is observed that the grain size distribution pattern of the tidal stretch of the Hooghly-Bhagirathi river system has a tendency of attaining a log-skew-Laplace distribution model whereas the other three non-tidal river sediments show a clear log-normal distribution pattern as a best-fit statistical model.

A possible reason has been attributed. During prolonged transport of bed material, the coarser and finer sediments are separated in the following manner: coarser sediments are buried below the advancing ripple fronts whereas the finer sediments are transported further downstream as suspension load. As a result, sediments of a narrow range of size classes are active in the distribution pattern giving rise to symmetric and unimodal distribution patterns which are the main criteria of log-normality. On the other hand, due to the upstream and downstream movement during sediment transportation in tidal reaches, the same sediments are constantly reworked and the segregation of different size fractions is hindered causing a log-skew-Laplace distribution pattern.

4.1 Introduction

A number of attempts have been made to discriminate different depositional sedimentary environments with varying degree of success using textural parameters (Mason & Folk, 1958, Shepard and Young, 1961, Folk, 1966, Friedman, 1961, 1967, 1979, Hails and Hoyt, 1969, Sahu, 1983, and many others). Sahu (1964), Moiola and Spencer (1979) used discriminant analysis and Klovan (1966) used factor analysis to differentiate depositional sedimentary environments. As each environment is related to a particular hydrodynamic condition of deposition, neither assumption is wholly valid. The idea of Bagnold (1937) highlighted the demerit of conventional grain size distribution diagrams where the grain size frequency is plotted against log particle size (phi scale). It is rather interesting that grain size distribution patterns show different styles when both size and grain size frequency in each size class are plotted in log-log scale to differentiate different depositional environments (Bagnold and Barndorff-Nielsen, 1980, Barndorff-Nielsen *et al.*, 1982, Fieller *et al.*, 1984, Vincent, 1986, Fieller and Flenley, 1992, Purkait and Mazumder, 2000, Purkait, 2000, 2002, 2003, 2006). In the present study, stress was given on the dynamic processes operating in the fluvial environments of different geomorphic setup leading to the development specific style of size distribution pattern. Sediment samples were collected from the cross-bedding foresets covering different stretches of five rivers– Hooghly-Bhagirathi, Ajoy, Damodar and the Usri (Fig. 4.1). The different samples from these rivers are taken to be independent of each other, and their spatial distribution is not considered. The effect of this sampling mechanism has also not been studied.

4.2 River System Characteristics

4.2.1 Ganga-Bhagirathi River System

The Ganga-Bhagirathi-Hooghly and their tributaries form the main deltaic river system of the Ganga belt lying in the state of West Bengal.

The Bhagirathi, a major spill distributary of the Ganga, branches off from the right bank of the Ganga at about 42 km downstream of Farakka. The lower reaches of the Bhagirathi-Hooghly river system down stream of Nabadwip-Swarupganj is known as the Hooghly river, while the upper reaches of the river system upstream of Nabadwip-Swarupganj upto its off-take from the Ganga is called Bhagirathi. The lower reaches of the river is tidal in nature. Previously about 300 years ago the tidal limit was nearly 16 km downstream of Nabadwip. Now-a-days, the tidal influence moved further north of about 8 km upstream of Nabadwip. The Hooghly empties into the Bay of Bengal at Sagar after traversing about 282 km.

A prominent mid-channel bar, about 2 km in length developed near Halisahar, and a point bar formed near Garifa, have been examined for collection of sediment samples. The natural levee deposits and floodplain sediments occurring around Kalyani and Majherchar have also been sampled for grain size analysis. Sediment samples have been collected from the stream bed near Anandabas, Swarupganj, Nabadwip, Udainarayanpur, Katwa, Berhampore and Jiaganj.

4.2.2 Ajoy River

The Ajoy River, originating in the Santhal Parganas Hills joins the Bhagirathi near Katwa and its confluence with the Ganges is about 220 km north of the mouth of the Ganges where the Ganges meet the Bay of Bengal. The Ajoy River is about 288 km long with a catchment area of about 6760 km^2 out of which less than 50% is hilly and the rest is flat plains. This flashy tributary causes havoc as it debouches into the main stream at Katwa, where the capacity of the latter is only about 1, 27,000 cusecs (Bhattacharya, 1973). Floods have occurred on a number of occasions. Flood waves accompanied by spasmodic bed movement must have followed the flood waters right into the navigation channel of the Hooghly beyond the Calcutta Port. The river does not experience any tidal influence whatsoever. The river is characterized by seasonal flash floods during the monsoon (Bhattacharya, 1973). Samples were collected from a lateral bar of the Ajoy River downstream of the bridge over it near Ilambazar.

A total of 20 sediment samples were collected from left lateral bar just downstream of the bridge. Samples were collected from the upper 2 cm of the toe of the valley water of the water line.

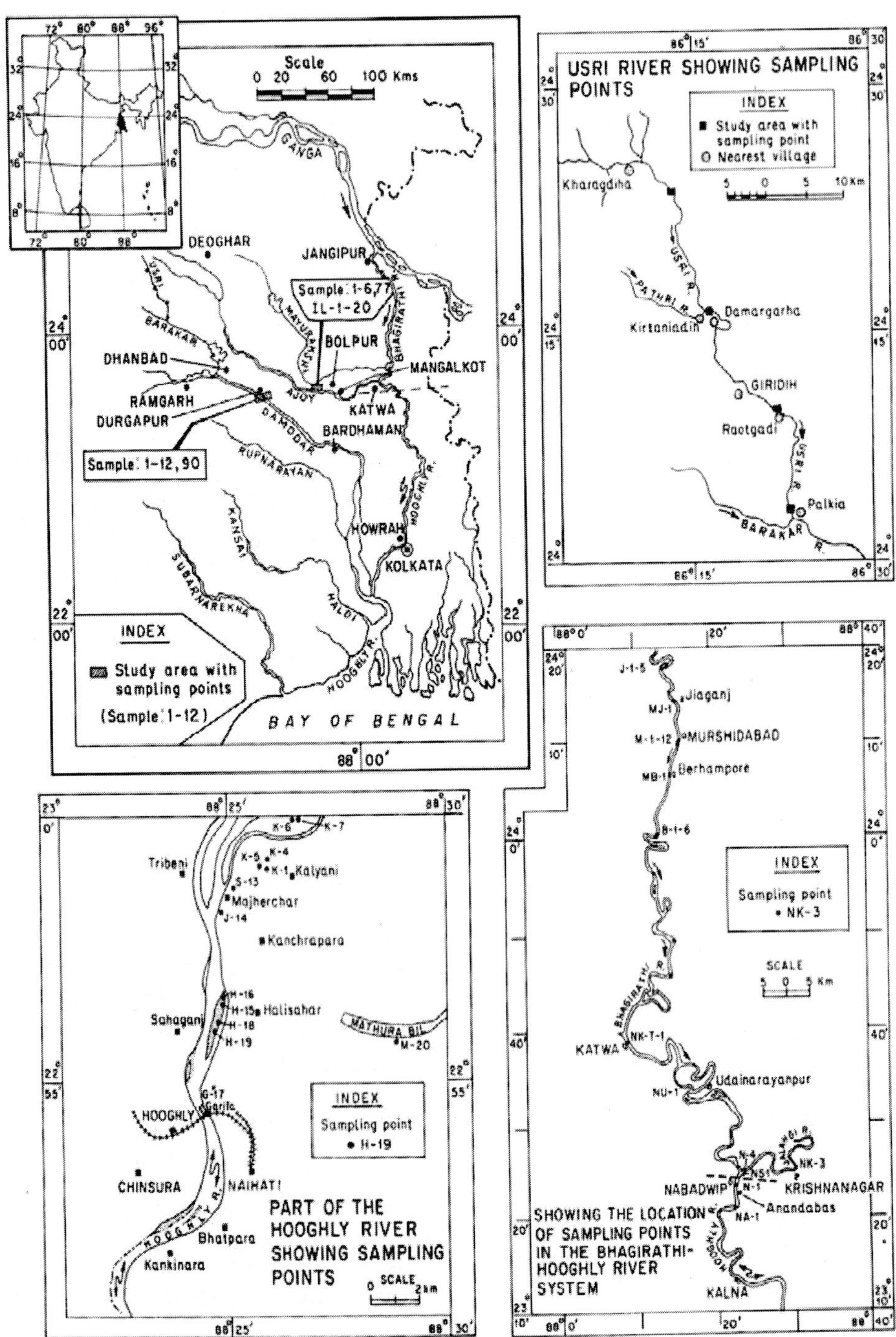

Fig. 4.1 : Location map showing the river systems and sampling points.

4.2.3 Damodar River

The Damodar River, circumscribing parts of Bihar and West Bengal, originates from the hills at an elevation of about 590 m in the state of Bihar. It is a small rain-fed river, 541 km long, originating from the Khamerpet hill (1068 m), near the trijunction of Palamu, Ranchi and Hazaribag districts of Jharkhand. It flows through the cities, Ramgarh, Dhanbad, Asansol, Durgapur, Bardhaman and Howrah before ultimately joining the lower Ganga (Hooghly estuary) at Shyampur, 55 km downstream of Howrah. The river is fed by a number of tributaries at different reaches, the principal ones being Jamunia, Bokaro, Konar and Barakar. The total catchment area of the basin is about 23,170 km^2, of this, three fourths of the basin lies in Jharkhand and one-fourth in West Bengal. The major part of the rainfall (82%) occurs during the monsoon season (June-August) with a few sporadic rains in winter. Broadly, the geology of the lower Damodar basin is quite distinctive from the upper basin. The lower basin is more or less wholly covered with a thick veneer of alluvium over the rocks of Tertiary age, while the upper basin is characterized by an abundance of granite gneisses of Archaean age and sandstones, shales and coal seams of Gondwana age. Damodar basin is an important coal bearing area and at least seven coal fields are located in this region.

The river generally flows in a south-easterly direction for about 290 km. After entering the deltaic plain, the river abruptly changes its course to a southerly direction and flows into the Hooghly River (a distributary of the Ganges-Brahmaputra River system). The total length of the river is about 535 km. The principal tributary, the Barakar, joins the Damodar River from the north, just before it enters the deltaic plain. The slope of the river changes as it flows towards the deltaic plain. From the source to 240 km, the slope is 0.002; from 240 km to 395 km, the slope is 0.0006; and from 395 km to 535 km, the slope is 0.002. The total drainage area of the river is about 21500 km^2, of which 10750 km^2 is the drainage area of the upper Damodar River up to the confluence with the Barakar River.

4.2.4 Usri River

The Usri is a small, low sinuous river of 90 km long, characterized by fluctuating stream discharge. It is a third order tributary river of the Barakar, situated in the state of Jharkhand, eastern India. During summer months (April-May), the river dries up almost completely exposing the point bars and the stream bed. It originates in several small streamlets and headward eroded gullies in the source area. The source area displays badland topography with the development of mesas and buttes over highly erodible pebble-sand deposits that contribute to the sediment load of the Usri River. It passes through mainly alluvial sediments and at places over the gneissic country rocks. It maintains an almost uniform gradient of 1:664 except in the lower 8-9 km reach, where the gradient abruptly changes to 1:150. The maximum bankfull width varies from 75 to 604 m, bank full depth 1.30 to 3.02 m, stream bed slope 0.00125 to 0.00275, sinuosity 1.34 to 1.48, and bankfull discharge 210.40 to 351.91 m^3/s whereas the maximum mean annual discharge is 36.62 m^3/s. A total of 42 sediment samples from all along the foresets of different size of cross-beddings were collected for grain size analysis.

4.3 Statistical Modeling

Grain-size distribution patterns have been compared with three distribution models: Normal (Gaussian), Hyperbolic, and Skew-Laplace. Parameter estimates of the model distributions are given in table 4.1.

4.3.1 Normal (Gaussian)

The density function for normal distribution is: g (x; μ, σ) = 1/σ √(2π) exp {-1/2 $(x-\mu)^2/\sigma^2$ }, where σ = standard deviation and exceeds zero, m = mean size, and x = observed variable, π is a constant and equals 3.1428. The log probability function is a parabola. In the normal distribution, the mean, median and mode values coincide, the graphic skewness = 0, and the kurtosis = 1.

4.3.2 Hyperbolic

Denoting the size variable by f and the density function S (ϕ), the equation of the hyperbolic curve is of the form (Barndorff-Nielsen, 1977):

$$Log_e S(\phi) = \nu - 1/2(\gamma_1 + \gamma_2)\sqrt{\delta^2 + (\phi - \mu)^2} + 1/2(\gamma_1 - \gamma_2)(\phi - \mu)$$

Four parameters are utilised to calculate it. These are γ_1, γ_2, μ and δ where γ_1 and γ_2 represent the slopes of the left and right asymptotes of the hyperbola respectively, μ is the abscissa of the intersecting point of these two asymptotes, the ordinate *v* equals the logarithm of the norming constant, which is adjusted to agree with the observed frequency and is obtained by putting the value of μ in the equation of either of the two asymptotes for the corresponding ordinate values.

The scale parameter δ (>0) can be expressed as: $\delta = 2(\mu^* - \mu)\sqrt{(\gamma_1 \gamma_2)}/(\gamma_1 - \gamma_2)$, where μ* is the observed mode of the distribution. For computing skewness and kurtosis, the graphic method of Folk and Ward (1957) was used. In a symmetric distribution where $\gamma_1 = \gamma_2$, μ* also equals μ. Such a distribution may be described as log-normal if the graphic skewness (Sk_1) = 0 and kurtosis (K_G) =1 (Folk and Ward 1957).

4.3.3 Skew-Laplace

The log- skew-Laplace distribution can be written as:

$$G(x;\alpha,\beta,\mu) = (\alpha+\beta)^{-1}\exp\{(x-\mu)/\alpha\} \text{ for } x \le \mu \quad 1$$

$$= (\alpha+\beta)^{-1}\exp\{(x-\mu)/\beta\} \text{ for } x > \mu \quad 2$$

where x indicates the observed variable and α, β, μ are parameters. The α equals the slope of the left asymptote i.e. coarser fractions, the β equals the slope of the right asymptote i.e. finer fractions, the μ equals the point of intersection of the two asymptotes and corresponds to the particular grain size obtained from the X-axis where the size classes are shown.

The estimated values of α and β were calculated from equations (1) and (2). The parameter μ must be such that

$$\hat{\beta}/\left(\hat{\alpha}+\hat{\beta}\right) = \left(\sum_{x>\mu} \textit{weight frequency}\right)/\textit{Total weight frequency} \quad 3$$

where $\hat{\alpha}$ and $\hat{\beta}$ are estimated values of α and β respectively. The μ is a lower limit that is adjusted so that the right side equals the left side of the equation.

Lastly, to obtain a quantitative idea of the discrepancies, the weighted relative errors between the computed and the observed values were calculated (table 4.2) for these three distributions by the following formula (Ghosh *et al.*, 1986, Purkait, 2002):

Error (E) $=\sqrt{\Sigma[(S_C - S_0)^2/S_0 T]}$, where S_C is the computed values (i.e wt% of calculated amount in each sieve fraction) as derived by the equations of different distribution models, S_0 is the observed values (i.e wt% of measured amount in each sieve fraction) and T is the total value observed. Here T equals 100 as the frequencies are calculated in percent.

4.4 Results of Statistical Analysis

The average of all parameter estimates of three hypothesized model distributions is shown in table 4.1. The computed error (E) in three model distributions is shown in table 4.2.

Table 4.1 : Average of all parameter estimates of three hypothesized model distributions.

River	Log -normal		Log-hyperbolic				Log-skew-Laplace		
	$\mu_N(\phi)$	$\sigma(\phi)$	γ_1	γ_2	$\mu_H(\phi)$	$\delta(\phi)$	α	β	$\mu_L(\phi)$
Hooghly	2.724	0.374	4.755	4.200	3.123	-0.630	0.716	0.159	3.568
Bhagirathi	2.919	0.528	3.833	3.266	3.607	1.926	0.485	0.413	3.769
Ajoy	1.296	0.605	3.142	3.797	1.508	4.609	0.754	0.375	1.887
Damodar	1.790	0.474	4.645	5.086	2.057	-15.966	0.544	0.338	2.225
Usri	1.126	1.042	2.243	3.306	1.591	-0.254	0.824	0.546	1.531

Table 4.2 : Interpretation of the results of computed error (E) in three model distributions.

River	Nature of the river	Total samples	No. of times wins Log-normal	Log -hyperbolic	Log -skew -Laplace
Hooghly	Tidal	22	5	9	8
Bhagirathi	Tidal	23	3	6	14
Damodar	Non-tidal	13	6	4	3
Ajoy	Non-tidal	26	16	3	7
Usri	Non-tidal	42	19	10	13

4.5 Discussion

From the computed error (table 4.2), it is observed that out of 22 samples from the tidal stretch of the Hooghly River, log-hyperbolic distribution fits best in nine cases (40.9%), log-skew-Laplace fits best in eight cases (34.4%), and log-normal fits best in five cases (22.7%). Out of 23 samples from the tidal stretch of the Bhagirathi River, log-hyperbolic distribution fits best in six cases (26.1%), log-skew-Laplace fits best in 14 cases (60.9%) and log-normal fits best in three cases (13%).

In case of non-tidal river, out of 13 samples from the Damodar River, log-normal distribution fits best in six cases (46.2%), log-hyperbolic fits best in four cases (30.8%), and log-skew-Laplace fits best in three cases (23.1%). Out of 26 samples of the Ajoy River, log-normal distribution fits best in 16 cases (61.5%), log-hyperbolic fits best in three cases (11.5%), log-skew-Laplace fits best in seven cases (26.9%). Out of 42 samples of the Usri River, log-normal distribution fits best in 19 cases (45.2%), log-hyperbolic fits best in 10 cases (23.8%) and log-skew-Laplace fits best in 13 cases (31%).

4.6 Conclusions

The statistical analysis of the sediment samples of the five rivers indicates that grain size distribution pattern of the tidal stretch of the Hooghly-Bhagirathi River system has a tendency of attaining a log-skew-Laplace distribution model whereas for the other three non-tidal rivers the log-normal distribution pattern is suggestive of a best-fit statistical model.

An attempt has been made to interpret this grain size distribution pattern in the light of the hydrodynamic condition of the rivers. During prolonged transport of bed materials, the coarser and finer sediments are separated in the following manner: coarser sediments are arrested in front of the lee sides of the advancing ripples whereas the finer sediments are transported further downstream as suspension. As a result, sediments of a narrow range of size classes are active in the distribution pattern giving rise to symmetric and unimodal distribution patterns which are the main criteria of log-normality. Such a hypothesis of grain sorting due to selective entrainment is also similar to the observation of Sengupta *et al.* (1999). On the other hand, due to the upstream and downstream movement during sediment transportation in tidal reaches, the same sediments are constantly reworked and the segregation of different size fractions is hindered causing a log-skew-Laplace distribution pattern.

Acknowledgements

The author is grateful to the Director General, Geological Survey of India for publication of this research. He is also grateful to Ex-Professors Supriya Sengupta and J.K. Ghosh of the Indian Statistical Institute for encouragement and valuable suggestions. Constructive comments from the reviewer Prof. J.K. Ghosh, Purdue University, USA and the other anonymous referee greatly improved the manuscript.

Bibliography

Bagnold, R. A. (1937). The size-grading of sand by wind, *Proceeding of the Royal Society of London*, A 163, pp. 250-264.

Bagnold, R. A., and Barndorff-Nielsen, O. (1980). The pattern of natural size distributions, *Sedimentology*, 27, pp. 199-207.

Barndorff-Nielsen, O. (1977). Exponentially decreasing distributions for the logarithm of particle size, *Proceeding of the Royal Society of London*, A 353, pp. 401-419.

Barndorff-Nielsen, O., Dalsgaard, K., Halgreen, C., Kuhlman, H, MÖller, J. T., and Shou, G. (1982). Variations in particle size distribution over a small dune, *Sedimentology*, 29, pp. 53-65.

Bhattacharya, S. K. (1973). Deltaic activity of the Bhagirathi-Hooghly river System, *Journal of Waterways, Harbours and Coastal Engineering Division, ASCE*, 99, 69-87.

Fieller, N. R. J., and Flenley, E. C. (1992). Statistics of particle size data, *Applied Statistics*, 41, pp. 127-146.

Fieller, N. R. J., Gilbertson, D. D., and Olbricht, W. (1984). A new method for environmental analysis of particle size distribution data from shore sediments, *Nature*, 311, pp. 648-651.

Folk, R. L. (1966). A review of grain size parameters. *Sedimentology*, 6, pp. 73-93.

Folk, R. L., and Ward, W. C. (1957). Brazos river bar : a study in the significance of grain-size parameters, *Journal of Sedimentary Petrology*, 27, pp. 3-28.

Friedman, G. M. (1961). Distinction between dune, beach and river sands from their textural characteristics, *Journal of Sedimentary Petrology*, 31, pp. 514-529.

Friedman, G. M., (1967). Diagnostic processes and statistical parameters compared for size frequency distribution of beach and river sands. *Journal of Sedimentary Petrology*, 37, pp. 327-354.

Friedman, G. M. (1979). Differences in size distributions of populations of particles among sands of various origins, *Sedimentology*, 26(1), pp. 3-32.

Ghosh, J. K., Mazumder, B. S., Saha, M., and Sengupta, S. (1986). Deposition of sand by suspension currents. Experimental and theoretical studies, *Journal of Sedimentary Petrology*, 56, pp. 57-66.

Hails, J. R., and Hoyt, J. M. (1969). Significance and limitations of statistical parameters for distinguishing ancient and modern sedimentary environments of the lower Georgia coastal plain, *Journal of Sedimentary Petrology*, 39, pp. 559-580.

Klovan, J. E. (1966). The use of factor analysis in determining depositional environments from grain-size distributions, *Journal of Sedimentary Petrology*, 36, pp. 115-125.

Mason, C. C. and Folk, R. L. (1958). Differentiation of beach, dune and aeolian flat environments by size analysis, *Journal of Sedimentary Petrology*, 28, pp. 211-226.

Moiola, R. J., and Spencer, A. B. (1979). Differentiation of eolian deposits by discriminant analysis, in E. D. Mckee, (ed.), A study of global sand seas, *U. S. Geological Survey, Professional Paper*, 1052, pp. 53-58.

Purkait, B. (2000). Morphology and growth pattern of the Usri river point bars, *International Journal of Sediment Research*, 15, pp. 445-457.

Purkait, B. (2002). Patterns of grain size distribution in some point bars of the Usri river, *Indian Journal of Sedimentary Research*, 72 (3), pp. 367-375.

Purkait, B. (2003). Statistical Modelling of grain size distribution patterns of some point bars of a modern river Usri, India, *Proceedings of the Fourth South Asia Geological Congress, New Delhi, India (GEOSAS-IV), Geological Survey of India*, pp. 83-91.

Purkait, B. (2006). Grain-size distribution patterns of a point bar system in the Usri River, India, *Earth Surface Processes and Landforms*, 31, pp. 682-702.

Purkait, B., and Mazumder, B. S. (2000). Grain size distribution – a probabilistic model for Usri river sediments in India, in Z. Wang and S. Hu (eds.) Stochastic Hydraulics, A.A Balkema: Rotterdam, pp. 291-298.

Sahu, B. K. (1964). Depositional mechanisms from the size analysis of clastic sediments, *Journal of Sedimentary Petrology*, 34, pp. 73-84.

Sahu, B. K. (1983). Multigroup discrimination of depositional environments using size distribution statistics, *Indian Journal of Earth Sciences*, 10, pp. 20-29.

Sengupta, S., Das, S. S., and Maji, A. K. (1999). Sediment transportation and sorting processes in streams, *Indian National Science Academy, Proceedings*, 65A, pp. 167-206.

Shepard, F. P. and Young, R. (1961). Distinguishing between beach and dune sands, *Journal of Sedimentary Petrology*, 31, pp. 196-214.

Vincet, P. (1986). Differentiation of modern beach and coastal dune sands – a logistic regression approach using the parameters of the hyperbolic function, *Sedimentary Geology*, 49, pp. 167-176.

CHAPTER 5

Estimation of Shape Changes of Skull Roof Bones in *Benthosuchus sushkini*, a Temnospondyl Amphibian from the Triassic of Russia

Ritabrata Dutta, Parthasarathi Ghosh and Dhurjati P. Sengupta

Abstract

The study of size, shape and pattern of various bones in the skull of vertebrate fossils constitutes the foundation of their taxonomy and plays a role in understanding their evolution as well. Size and shape change during lifetime, and the growth of an individual bone involves both isotropic enlargement of its boundary due to uniform growth and differential enlargement that changes its shape. This paper attempts to develop a methodology for quantitative assessment of the pattern of ontogenic shape changes for all the individual skull roof bones of a taxon (*Benthosuchus*) of an extinct amphibian (temnospondyl) from the Lower Triassic of Russia. Published diagrams depicting skulls of different ontogenic stages have been used. A measure for the degree of shape changes is also proposed. In this study, the skull roof bones in dorsal view were represented as polygons. The boundary of each polygon of a juvenile skull was compared with that of an adult. For comparative purpose, the polygons of the younger skull were placed optimally inside that of the older skull, under the assumption of maximised uniform (isotropic) growth. The variance of the distances between the two boundaries measured at a number of points along the boundaries was considered as a measure for the degree of shape change. The results obtained are in partial concurrence with the published observations traditionally obtained from anatomical studies for the above mentioned growth series.

5.1 Introduction

The study of size and shape of the skulls of fossil vertebrates constitutes the foundation of their taxonomy. It also plays an important role in understanding the evolutionary pathways. The bones of a vertebrate grow in size and change in shape within their life span (Foote and Miller 2007). The rates at which these changes take place generally depend on factors like nutrition, growth stage, environmental conditions etc. The bones grow uniformly on all sides and become larger. They also grow differentially in different directions to change their shapes. In this paper, we present a technique for quantitative assessment of the shape changes through ontogeny of the skull roof bones of an extinct amphibian group (Temnospondyli).

Temnospondyls are very common in the Mesozoic non-marine deposits of the world and a number of well-illustrated descriptions are available in literature (Romer 1947, Schoch and Milner 2000). Their skulls were generally flat and that makes them suitable for 2D analysis in most cases. Moreover, for some taxa, data are available on juvenile and adult skulls from which a growth series can be reconstructed. *Benthosuchus sushkini* (Efremov, 1929, Bystrow and Efremov, 1940) is one of the best known taxa which depicts such a growth series (Schoch and Milner 2000, figure 28, p. 29).

Previously, the distance between a pair of skull roof parameters (like the pre or postorbital length, internarial width etc.) or their ratios were used to indicate growth patterns and shape changes of the temnospondyl skull bones (Colbert and Imbrie 1969, Welles and Cosgriff 1965, Welles and Estes 1969, Sengupta and Ghosh 1993). During the last decade or so, mathematical models have frequently been used in paleontological literature (Temple 1992) as well as in the study of the vertebrates (Orlov 1991). Quatitative methods are now being used for temnospondyls as well. Recently Sengupta *et al.* (2005) developed a detailed descriptor using a Fourier based method to quantify the temnospondyl skull outlines using 58 temnospondyl skulls that included many adults and a few juveniles. Subsequently, Stayton and Ruta (2006) used relative wraps to study the geometric morphometry of 62 species of stereospondyls. They used 21 landmarks to quantify the evolution of the different temnospondyl skulls. Thereafter, Witzmann and Scholz (2007) made a study of the growth stages of *Archegosaurus decheni* from a large data set of 96 specimens by using cartesian transformations, bivariate regression analysis and principal component analysis. The present study is, however, directed towards a quantitative assessment of ontogenic shape changes of the individual bones of a particular taxon. The skull outline is a collective expression of shape and organisation of the individual bones of the skull. Thus, during ontogenic changes, all the bones grow at various rates. In the present study, each of the skull roof bones of smaller individuals is optimally placed within the same bones of the successively larger individuals and the growth pattern is estimated from that. The present work aims to test the above technique on 30 skull roof bones of each of the three successive individuals of *Benthosuchus sushkini*.

5.2 Material and Method

5.2.1 Preparation of Digital Bone Boundary Dataset

Published illustrations of three skulls representing juvenile to adult specimens of *Benthosuchus sushkini* (Schoch and Milner, 2000, figure 28, p. 29) are used in this work. These illustrations deal with one of the best-preserved and often cited ontogenic growth series described among temnospondyls. The skull outline and the sutures on the dorsal view of the skulls are used in the present analysis (Fig. 5.1).

The illustrations of the skull roof bones were scanned using a flatbed scanner and the skull outline and sutures are manually digitized from the scanned image with the help of a computer drafting software (AutoCAD). The digitized features are corrected for scale and orientation and then, with the help of a GIS software (ArcGIS 8), a polygon feature class has been derived from the topology of the network of lines. The polygons are taken as representatives of individual bones or openings. A comparison of correlated polygonal boundaries between skulls was undertaken in the subsequent analysis.

It was assumed that the boundary of a bone would expand during ontogeny to give rise to the boundary at a more matured stage. Thus, if a bone B growing over time is represented by a series of bones B_1, B_2, ... B_t for t time points then the bone B_i will be contained inside the bone B_{i+1}. Now, if this expansion takes place uniformly for all the parts of the boundary, the shape will be preserved and the bone will only increase in size (uniform growth). Shape will change if the growth is differential (differential growth). To compare the degree of expansion in different directions, it is first necessary to place the B_i within B_{i+1} at a suitable location. As bone boundaries are represented by non-convex polygons, one simple way to place a smaller bone within the larger one is to match them by their respective centroid. However, centroid is a geometric landmark. Its location is determined by the shape of the polygonal boundary. Thus, the centroid of B_i may not always coincide with the centroid of B_{i+1} after growth (Fig. 5.2). In fact, it was noted that for many polygon pairs, the B_{i+1} does not fully contain B_i when they are matched against their centroid, though it was possible to place B_i completely within B_{i+1} at other locations.

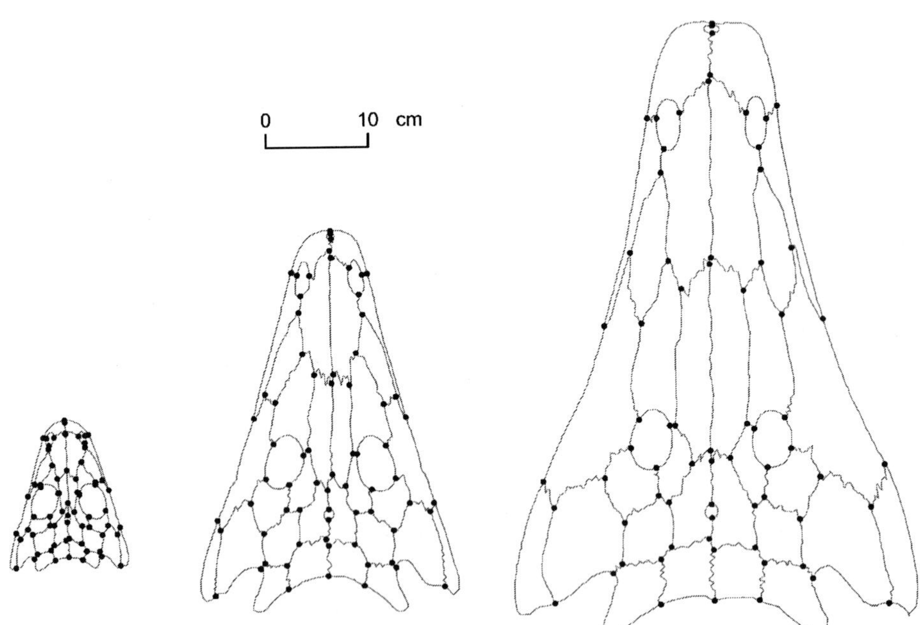

Fig. 5.1 : Juvenile, sub-adult and adult skulls (dorsal view) of *Benthosuchus sushkini*, after Schoch and Milner (2000). Outlines of the individual skull bones and openings are shown along with the "node points" where sutures meet.

This work, therefore, aims first to locate the optimal placement of B_i inside B_{i+1} so that the component of uniform (isotropic) growth is maximized. That is to find where the largest copy of the smaller bone could be placed inside the larger one. However, since an individual bone is a part of a mosaic that constitutes a skull, some constraints are to be imposed while placing the smaller bone inside the larger one.

A scrutiny of skull bone organization suggests that the topology of the bone polygons remains invariant during growth. In other words, the bones that share common boundaries in a juvenile skull remain adjacent during growth. Therefore, the points lying on the bone boundary where three adjacent bones meet (say nodes), can be used to constrain placement of a juvenile bone within its adult counterpart and can be taken as references for one to one map between corresponding bones of a series (Fig. 5.1). These reference points may be used to divide the polygonal boundary into a number of connected segments (Fig. 5.3). As the end points of these segments of B_i are in a one to one map with the corresponding end points on B_{i+1}, the segments will also be in one to one map with the segments of the larger bones.

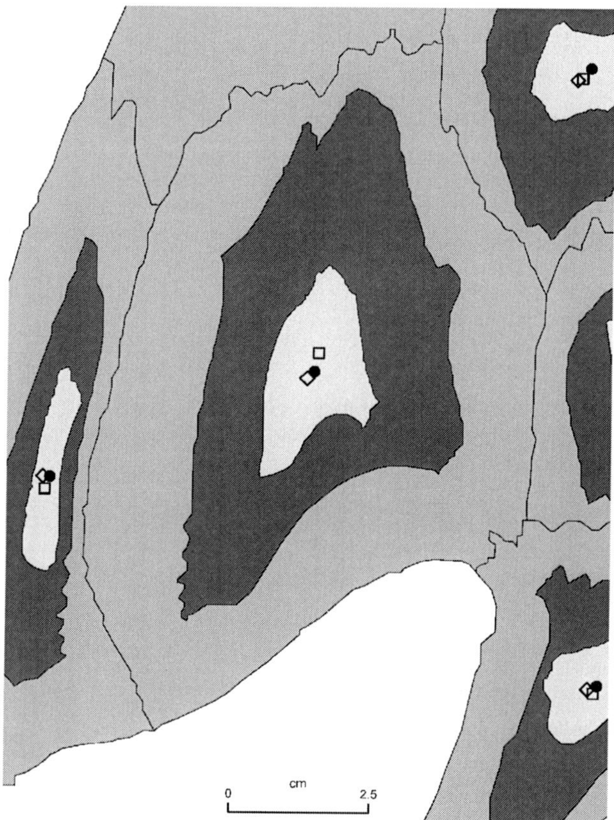

Fig. 5.2 : Left posterolateral corner of the skull of adult *B. sushkini* with corresponding bones of juvenile and sub-adult skulls placed optimally within it. The centroids of bone polygons of the adult, sub-adult and juvenile skulls are shown as open diamonds, open squares, and solid dots respectively. Note that at optimal placement the centroids of corresponding bones are not coincident.

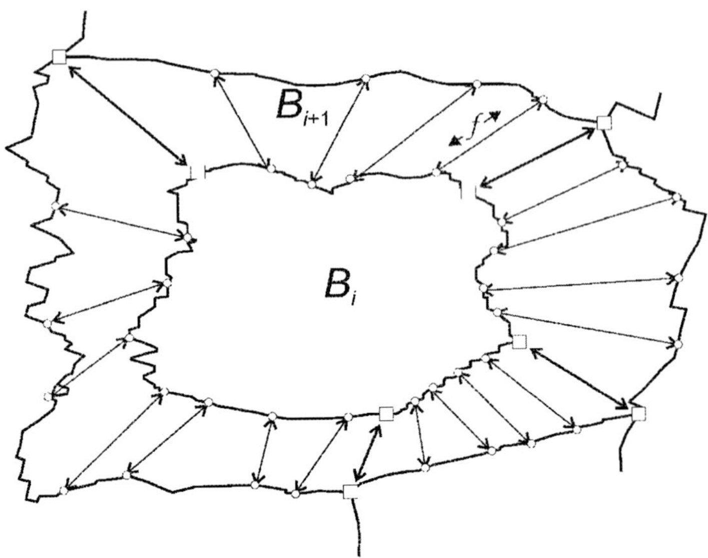

Fig. 5.3 : Diagram showing the correspondence (shown by the double ended arrows) between the points (dots) lying on the polygonal boundaries of a bone of the younger skull (B_i) and those on a more matured skull (B_{i+1}) at the optimal placement. The reference points lying at the junction of three sutures that were used to subdivide the boundaries into a number of segments are shown by open squares.

Out of 36 polygons representing bones and openings of studied *Benthosuchus shushkini* skulls, sixteen polygons have five to six nodes on their boundary. 8 bones have 8 nodes and 6 bones have 4 nodes. Only 6 polygons have less than 4 nodes and they correspond to the quadratojugals (each having 3 nodes), the pineal foramen and the anterior fenestrae. Each of the two openings has 2 nodes on their boundary. For this study, we consider only the individual bone boundaries that are constituted of four segments, as it is easier to generate a boundary-conforming grid subsequently used for optimization. Therefore, four out of the available nodes on the bone boundary are chosen in such a way that the boundary is segmented into four roughly equal parts. For the polygons having three nodes one additional node was inserted: at the tip of the wings for the quadratojugal and on inner part of the nasal opening. Two nodes were inserted for each of the two medial openings so as to divide its boundary into segments of roughly equal lengths. A fixed number of points (N), along the boundary have been taken for each boundary segments. A large N ($\geq$ 200) approximates the boundaries with reasonable accuracy. Therefore, if there are M segments in a boundary then the total number of points on the boundary will be M x N. For the bones in a series, the number of segments remains constant (i.e., $M = 4$) and the number of data points per segment (i.e., N) is also made fixed.

5.2.2 Formulation of Optimization

The optimization aims to minimize the criterion D over the domain of all possible placements of B_i inside B_{i+1}. The criterion D is defined as,

$$D = \sum w_j f_j^2$$

Where, $\{f_j\}_{j=1,M\times N}$ are the linear distances between the M x N pair of points on B_i and B_{i+1} respectively that are related by one to one map and $\{w_j\}_{j=1,M\times N}$ is the sum of arc lengths between two adjacent points to its left and right measured on B_{i+1}. The segments of a boundary are of unequal lengths, though all of them are represented in this study with equal number of points (i.e., N). The weightage factor was incorporated to compensate the unequal arc lengths being represented by equal number of points.

The lines that connect the M x N pairs of points on the segments of the two polygonal boundaries (i.e., B_i and B_{i+1}) are radially disposed and f_i represents their lengths. Therefore, a placement that gives the lowest value of D allows placing the largest copy of B_i inside B_{i+1}. In other words, the minimum D identifies the placement where the component of uniform growth can be maximized.

However, the optimisation process was so constrained that

1) One to one mapping between the segments of B_i and B_{i+1} is preserved, i.e., the placements should be such that, no linear paths (connecting the corresponding point on two bones), will intersect each other.

2) B_i should be contained in B_{i+1} as completely as possible. As the bones used for this analysis do not belong to the same individual, for reasons obvious, it has been noted that for some pairs of bones, B_i is not completely contained within B_{i+1} for any placement.

5.2.3 Method for Finding the Optimal Placement

In most cases the bones under study are non-convex polygonal objects. Finding the domain of all placements considering translation and rotation of one non-convex polygon inside another is still an unsolved geometrical problem (Agarwal *et al.*, 1999). The same authors recently proposed a suitable algorithm, for placement of a convex polygon inside a polygonal environment. But the placement of largest homothetic copy of a convex polygon has been solved only by Megiddo's parametric searching technique (Agarwal *et al.*, 1999). It should be noted that the present problem is not identical, as it needs to solve an optimization problem considering two non-identical, non-convex polygons. An algorithm for finding the optimal solution for this problem is not yet available. Therefore, instead of trying to find an exact mathematical solution, an approximate solution by adopting a computational technique has been attempted here.

In this work any internal point of B_i (say the centroid without any loss of generality) was placed at each point of a dense boundary conforming grid inside B_{i+1}. For each such placement, B_i was rotated around its centroid within the limits of constraints mentioned above. The values of D were computed for each placement and for all rotations, if permitted by the constraints mentioned above. The Cartesian coordinates of the nodes and the angle of rotation that corresponds to the minimum value of D were considered for the optimal placement of B_i inside B_{i+1}.

5.2.4 Measure for the degree of shape change

The set of f_i at the optimal placement of B_i in B_{i+1} provides the degree of expansion of the bone boundary in different directions. If all the *fs* are exactly equal then the expansion is isotropic i.e., the bone simply got enlarged and the differential growth component should be equal to zero. On the other hand, a large variance of f would suggest a dominance of differential growth (i.e., change in shape). Therefore, the variance of f provides a measure for the degree of shape change during the growth. However, the nature of shape change cannot be determined from the variance values only. The orientation of the axial vectors means, for the vectors connecting the corresponding boundary points of B_i and B_{i+1}, provides additional information regarding the direction of overall elongation of the individual bones during growth.

5.3 Results and Discussion

5.3.1 Results

Figure 5.4 shows the optimal placements of the bones and openings of different individuals of *B. sushkini*. The results suggest that the bones (and openings) of smaller (juvenile) individuals can be well accommodated within the bones (and openings, respectively) of successive larger (older) individuals. Figure 5.5A shows the variances values and axial vector means between a juvenile and an adult skull. Variance values and axial vector means between skulls of juvenile and a sub-adult, and a sub-adult and an adult have also been included to demonstrate the changes related to the different stages of ontogeny (Fig. 5.5B & C). The degree of shape change for the paired bones (i.e., bones disposed symmetrically from the medial line) is roughly similar in most of the cases. This is especially obvious for the later stage of growth.

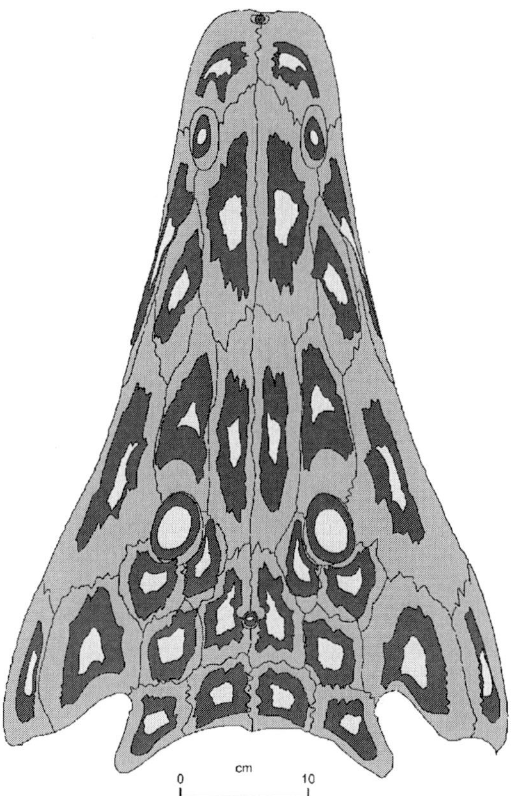

Fig. 5.4 : Optimal placements of the corresponding bones of the juvenile skull (White) inside the sub-adult skull (Black) and also that inside the adult skull (Grey).

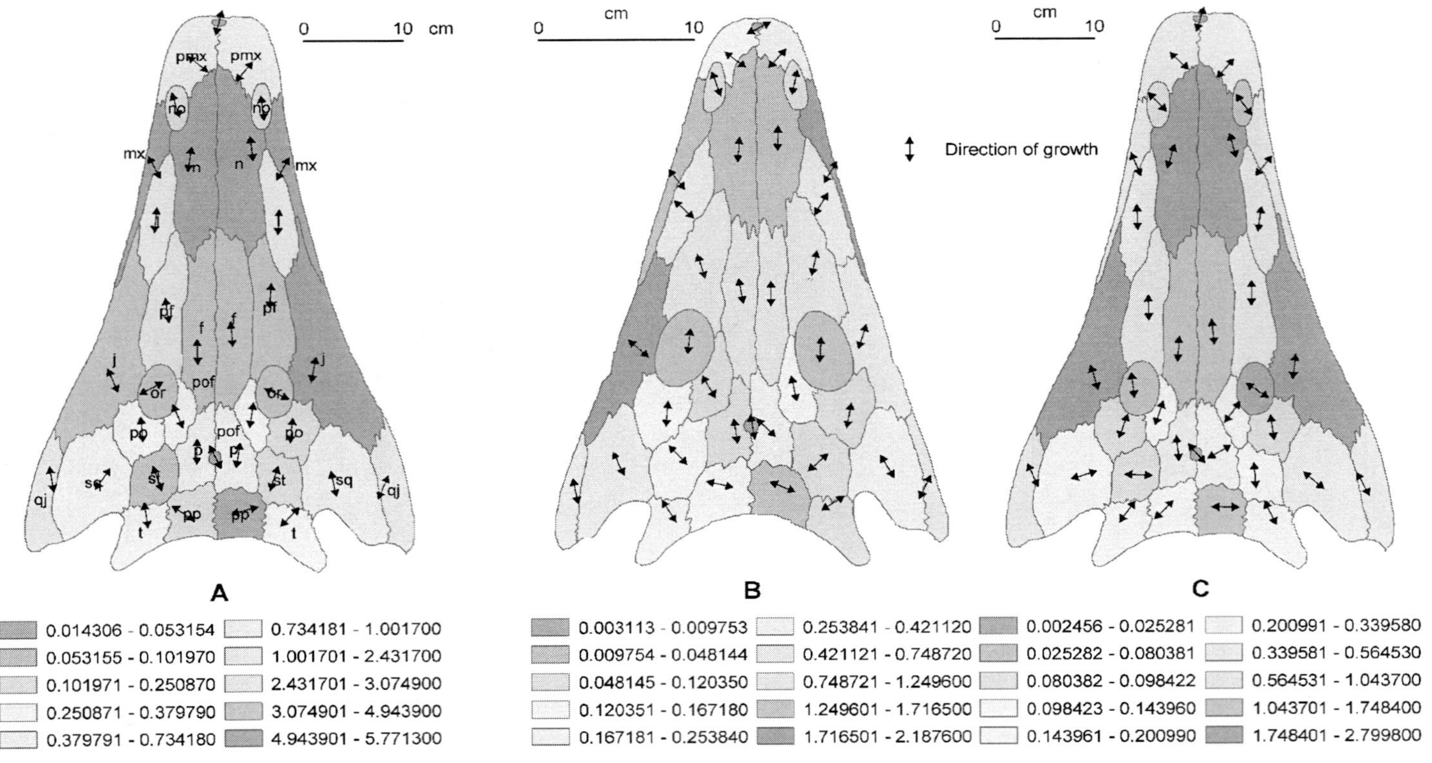

Fig. 5.5 : Pattern (and corresponding variance values and axial vector means) of shape changes of the individual skull bones / openings during the growth of *B. sushkini*. A) Between a juvenile and an adult skull; B) Between a juvenile and a sub-adult skull; C) Between a sub-adult and an adult skull. Different colours represent the degree of shape changes of individual bones. The overall direction of elongation of individual bones is shown by double-ended arrow. The abbreviated names of the bones and openings are indicated (see Table 5.1 for details).

On the basis of degree of shape change, the skull of *B. sushkini* can be subdivided into two distinct regions that are characterized by low and high shape changes. The postorbital bones (i.e., postparietal, supratemporal, postorbital, postfrontal) and those in the anterior tip of the snout (i.e., premaxillae) show medium to low shape changes during the whole ontogeny. The region in between the orbits and the nasal openings (i.e., the jugal, nasal, maxialla, frontal, prefrontal, and lachrymal) is characterized by high shape changes. The most dramatic changes occur for maxillae, jugals and nasals. It is interesting to note that the more intense shape changes of the maxilla take place in the postlarval stages of the growth (between the juvenile and the sub-adult stage). In contrast to pre-frontals, frontals undergo more intense shape changes in the later part of the ontogeny.

In general, the elongation of the component bones characterizes the shape modifications between the orbits and the nostrils. While the frontals, pre-frontals and lachrymals become dominantly elongate, the jugals and nasals show a component of widening along with elongation. Considering the above observations, it can be suggested that the ontogenic change from a convex to a concave lateral margin of the skull may be linked with 1) elongation of the mid-skull bones, and 2) uniform growth of both the postorbital and the anterior-most bones.

5.3.2 Discussion

The pattern of ontogenic changes as suggested by the results of this study are in congruence with the observations made by the traditional anatomical studies on the growth of *Benthosuchus* (Bystrow and Efremov 1940, Schoch and Milner 2000) in a broad sense. Comparison of the variance values of skull roofs bones of *B. sushkini* indicates that the jugals, nasals, frontal and the maxillae have undergone a high degree of changes from juvenile to adult (Table 5.1). Schoch and Milner (2000) also noted similar changes in jugal and premaxilla through ontogeny. However, they indicated that those bones had undergone a proportional change to fit in the larger size with ontogeny. The present study therefore suggests *contra* Schoch & Milner (2000) that 1) the variations of the skull roof bones through ontogeny are not (only) proportional changes, but also shape changes as well; and 2) the premaxillae do not undergo strong changes relatively to the other bones. The directions of the axial vector (Table 5.1) of the above mentioned bones indicate the directions in which the changes occur through ontogeny. Again, from Table 5.1, it is clear that the areas immediately anterior to the orbits of the chosen skulls show the maximum growth. The skulls clearly show progressive elongation of the snout when studied qualitatively. This phenomenon is also captured by our method (Fig. 5.5). Incidentally, many paleontologists including recent workers like Schoch and Milner (2000), Damiani (2001), Steyer (2003), and Stayton and Ruta (2006) have also observed that late Early Triassic and Middle Triassic stereospondyls show a tendency towards preorbital elongations (longirostry). Steyer (2002) noted that *Wantzosaurus elongatus*, Lehman 1961, a trematosaur, developed shorter posterior and longer anterior part through ontogeny. Incidentally, Steyer (2003) attempted a thorough study of *Watsonisuchus madagascariensis* (new combination, including *B. madagascariensis*, Lehman 1961) and noted that the subtriangular juvenile skulls of *B. sushkini* and *W. madagascariensis* are similar in shape and the later shows an ontogenic growth from semirostry to longirostry. However, in *Watsonisuchus* the longirostry is marked by prefrontal and frontals (Steyer 2003). In addition, there are examples of lachrymals growing larger through ontogeny (as in Permian *Cheliderpeton*, Wernerburg and Steyer 2002). In *B. sushkini,* as stated above, prefrontal, frontal, nasal, maxilla and jugals, are the bones that went through major shape changes. There are certain traits that are noticed through the ontogeny of *B. sushkini*. From juvenile to subadult the frontal, nasal and jugal bones had undergone maximum shape change. The jugals show lateral or diagonal growth directions while the nasal and frontal show anterior elongation. From subadult to adult, nasal, frontal, jugal, prefrontal and maxila are the major bones that show change in their shapes and sizes. A comparison between the skull roof bones of the juvenile and the adult stages indicates that the nasals, maxilae and jugals are the bones where maximum shape change occurred. Hence, apart from simple longirostry, a side-wise growth is also noted, particularly in the bones that border the skull outline through ontogeny. Interestingly, that also includes the premaxilae, which, despite the general longirostry, maintained an oblique growth.

Table 5.1 : Comparison of the directions of axial vector means in degrees and the variance values (measure of shape changes) of each bone of the skull roof of the juvenile, subadultand adult skull roofs of *Benthosuchus sushkini.*

Abbreviations used	Bone Name	Side of the skull	Comparison between the juvenile and the adult skulls		Comparison between the Intermediate and the adult skulls		Comparison between the juvenile and the intermediate skulls	
			Direction of axial Vector Means in degrees	Variance values (measure of shape changes)	Direction of axial Vector Means in degrees	Variance values (measure of shape changes)	Direction of axial Vector Means in degrees	Variance values (measure of shape changes)
F	Frontal	R	175.4800	4.6758	174.2700	1.3059	-1.7998	1.1706
		L	178.1700	4.3475	-1.7610	1.7484	1.6980	1.0235
Fo	Opening	M	13.8630	0.0238	16.1180	0.0122	62.0760	0.0031
J	Jugal	R	-1.6990	5.7052	-1.7320	2.7998	-1.6300	0.7189
		L	157.8400	4.9439	163.8300	2.2473	1.3311	2.0394
L	Lachrymal	R	-1.7710	2.1310	-1.7200	1.0437	-1.5124	0.7487
		L	-1.7450	2.1657	178.0200	0.5548	1.3465	0.5733
Mx	Maxilla	R	-1.5350	5.7713	-1.4330	0.8874	-1.5176	2.1876
		L	154.3200	5.6990	158.8800	0.5645	1.4355	1.7165
N	Nasal	R	173.2700	5.7265	166.2200	2.2498	-1.7895	1.4395
		L	-1.7180	5.2873	-1.6280	2.0608	-1.7590	1.4427
No	Nostril	R	172.8800	0.2096	145.7600	0.0804	-1.6880	0.0850
		L	163.8700	0.1702	138.2700	0.0984	1.6113	0.1042
Or	Orbit	R	115.4600	0.1020	125.0900	0.0253	-1.7684	0.0455
		L	-1.2040	0.0797	173.1400	0.0607	5.8981	0.0269
P	Parietal	R	-1.7180	0.6985	-1.2110	0.2010	1.3704	0.2538
		L	-1.7910	0.5214	174.5100	0.2706	1.7260	0.1070
Pf	Prefrontal	R	-1.7590	3.9710	-1.8000	0.7556	-1.6816	0.9957
		L	172.0600	3.0749	179.0200	0.8810	1.6314	1.0137
Pinf	Pineal Foramen	M	155.5000	0.0143	141.2700	0.0025	1.7089	0.0098
Pmx	Premaxilla	R	-1.4240	1.0017	-1.4110	0.3396	-1.4042	0.3893
		L	131.3200	0.8784	135.4700	0.4760	1.2930	0.1490
Po	Postorbital	R	-1.7390	0.2152	172.7000	0.0926	-1.6994	0.0866
		L	175.5100	0.3036	-1.6110	0.0850	-1.7516	0.1672
Pof	Postfrontal	R	-1.7140	0.3081	-1.4560	0.1594	1.6865	0.1369
		L	158.4300	0.3047	-1.6280	0.1100	1.5067	0.0920
Pp	Postparietal	R	74.6660	0.0532	91.4810	0.0798	-6.6909	0.0481
		L	-54.4420	0.2268	-1.3580	0.1102	1.0297	0.1660
Qj	Quadratojugal	R	-1.5720	2.4317	152.9500	0.4838	-1.5551	1.2496
		L	171.1100	2.1054	156.3600	0.7224	1.6897	0.3977
Sq	Squamosal	R	-11.1320	0.7342	-49.4560	0.2866	1.5044	0.1955
		L	-1.4580	0.2945	-1.0770	0.1096	1.5512	0.4211
St	Supratemporal	R	-1.6220	0.2509	175.6200	0.1440	-1.3287	0.0868
		L	161.4900	0.0815	-87.5690	0.0875	1.3806	0.1574
T	Tabular	R	-1.3690	0.3798	-25.1410	0.1126	-1.2451	0.1204
		L	171.3300	0.3234	34.4550	0.1289	1.4868	0.2263

Considering the complex geometry of the bones and the fact that the skulls of different individuals have been used to approximate the ontogenic growth, the present method provides reasonably sensitive measures for the shape change. Figure 5.5 indicates that the shape changes occurred in both sides of the skulls, i.e. in a more or less symmetric manner. However, the lack of perfect bilateral symmetry in shape changes was possibly induced by the use of skulls of different individuals in a reconstructed growth series. It can also partially be the artifacts of biological or taphonomic loss of symmetry in a single individual.

A visual comparison between the bones at different stages of growth suggests that various types of shape changes occur during ontogeny. Some of the bones like the squamosal, prefrontal, postorbital and nasal become somewhat triangular or bean like in shape; skull openings like orbits and nares become more elliptical; and for some like the jugal, the suture becomes more jagged. Our methodology, however, does not take into account the type of shape change. For example, in some specimens, the sutures marking the bone boundaries become increasingly sinuous through ontogeny. Therefore, even when the overall shape of a growing bone remains fairly constant, the variance calculated by the present methodology may increase due to increasing irregularity of the bone boundary. Further study of the directional properties of the vectors linking the boundary segments of the bones of juvenile and adult specimens can shade light to this aspect of ontogenic shape changes.

5.4 Conclusions

A new method to quantify the degree of shape changes during ontogenic growth of a temnospondyl taxon has been developed in this work. The results are in general agreement with the observations traditionally obtained from anatomical, qualitative studies. Moreover, this study estimates the changes associated with different phases of growth for individual skull roof bones in detail. There is marked elongation of the preorbital bones of *B. sushkini*. The skulls, through ontogeny, had changed from semirostry to longirostry. Apart from longirostry, some bones bordering the lateral outlines also show an oblique growth with reference to the orientation of the midline suture. Longirostry is common in many temnospondyl groups particularly in the trematosaurids and in the archegosauids. *Archegosaurus decheni* shows a negative allometry through ontogeny at the cheek region and hence had lesser amount of adductor muscles and their bite force was lesser than the parabolic skull bearing temnospondyls (Witzmann and Scholz 2007). Schoch and Milner (2000) said that the 'long-skulled stereospondyls' could deal with larger preys and the skull construction and musculature used to play an important role in this regard. Hence, longirostry was an important constructional aspect because adductor muscles are supposed to be less in the long-snouted forms. *Archegosaurus* has been considered as a sister group of sterospondyls (Yates and warren 2006) and their longirostry developed as a convergence to the piscivorous diet (Steyer 2003). On the other hand sequences of neurocranial ossifications of *Archegosaurus decheni* and *B. sushkini* are similar (Witzmann 2006). Like *Archegosaurus*, longirostry of *B. sushkini* was possibly related to their feeding habits.

For the semicircular or parabolic, deep skull bearing temnospondyls, the feeding mechanism is different. There is no longirostry associated with their growth. Though not attempted here, the present method could be useful to quantify the degree of ontogenic shape changes in short-faced temnospondyl skulls as well.

Acknowledgements

The Indian Statistical Institute, Kolkata, India, provided the computational facility required for this work. Authors thank Asit Saha of the Geological Survey of India, Kolkata, for his suggestions regarding GIS techniques used in this work. Prof. T. Roy Chowudhry, former Professor of the Indian Statistical Institute, Kolkata, India, was the source of constant inspiration behind this work.

Bibliography

Agarwal, P. K. Aronov, B. and Sharir, M. (1999). Motion Planning for a Convex Polygon in a Polygonal Environment, *Discrete and Computational Geometry*, 22, pp. 201-221.

Colbert, E. H. and J. Imbrie. (1956). Triassic metoposaurid amphibians. *Bulletin of the American Museum of Natural History.*, 110, pp. 399-452.

Damiani, R. J. (2001). A systematic revision and phylogenetic analysis of Triassic mastodonsauroids (Temnospondyli, Stereospondyli), *Zool. Jour of the Linn. Soc.*, 133, pp 379-499.

Bystrow, A. P. and J. A. Efremov. (1940). *Benthosuchus sushkini.* A Labyrinthodont from the Eotriassic of Sharzhenga River. *Travaux de l'Institut de Paléontologie de l'Académie des Sciences d'URSS*, 10, pp 1-119.

Efremov, I. A. (1929). *Benthosuchus sushkini*, ein neuer Labyrinthodont der permotriasischen Ablagerungen des Scharschenga Flusses, Nord-Dna Gouverment, *Academy of Science Bulletin, URSS*, Cl.1929, pp.757-770.

Foote, M. and Miller, A. I. (2007). Principles of Paleontology, Freeman and Company, New York, 354 p.

Lehman, J.P. (1961). Les Stegocephales de Madagascar, *Annal De Paleontologie*, 47, pp. 111-154.

Orlov, A. N. (1991). A mathematical method of studying vertebrate, *Paleontological Journal*, 2, pp. 113-118.

Romer , A. S. (1947). Review of the Labyrinthodontia. *Bulletin of the Museum of the Comparative Zoology, Harvard*, 99, pp. 1-367.

Schoch, R. R. and Milner, A. R. (2000). Stereospondyli, pp.1-203 *in:* P. Wellnhofer (ed.) Encyclopedia of Paleo-herpetology, 3b, 203p.,Verlag. München.

Sengupta, D. P. and Ghosh, D. P. (1993). Morphometrics of some Triassic temnospondyls, in S. G. Lucas and M. Morales (eds.) The Nonmarine Triassic. New Mexico Museum of Natural History and Science Bulletin, 3. pp. 423-428.

Sengupta, D. P., Sengupta, D and Ghosh, P. (2005). Bilaterally symmetric Fourier Approximation for the skull outline of temnospondyls and their bearing on shape comparison, *Journal of Bioscience*, 30(3), pp.377- 390.

Steyer, J. S. (2002). The first articulated trematosaur 'amphibian' from the lower Triassic of Madagascar: implication of the phylogeny of the group. *Palaeontology*, 45(4), pp. 771-794.

Steyer. J. S. (2003). A revision of the early Triassic "Capitosaurus" (Stegocephali, Stereospondyli) from Madagascar, with remark on their comparative ontogeny. *Journal of Vertebrate Paleontology*, 23(3), pp. 544-555.

Stayton, T. C. and Ruta, M. (2006). Geometric morphometrics of the skull roof of stereospondyls (Amphibia: Temnospondyli), *Palaeontology*, 49(2), pp. 307-337.

Temple. J. T. 1992. The process of quantitative methods in paleontology, *Palaeontology*, 35(2), pp. 475- 484.

Welles, S. P. and Cosgriff, J. W. (1965). A revision of the labyrinthodont family Capitosauridae and a description of *Parotosaurus peabodyi*, n. sp. from the Wupatki Member of the Moenkopi Formation of Northern Arizona, *University of California Publications in Geological Sciences*, 54, pp.1- 148.

Welles, S. P. and R. Estes. (1969). *Hadrokkosaurus bradyi* from the upper Moenkopi Formation of Arizona, with a review of the brachyopid labyrinthodonts. *University of California Publications in Geological Sciences*, 84: pp.1-56.

Werneburg, R. and Steyer, J. S. (2002). Revision of *Cheliderpeton vranyi* Fritsch, 1877 (Amphibia, Temnospondyli) from the Lower Permian of Bohemia (Czech Republic). Paläontologische Zeitschrift, 76 (1), pp. 149-162.

Witzmann, F. (2006). Devopmental pattern and ossification Sequence in the Permo-Carboniferous temnospondyl *Archegosaurus decheni* (Saar Nahe Basin, Germany), *Journal of Vertebrate Paleontology*, 26(1), pp.7-17.

Witzmann, F. and Scholz, H. (2007). Morphometric study of allometric skull growth in the temnospondyl *Archegosaurus decheni* from the Permian/ Carboniferous of Germany, *Geobios*, 40(4), pp. 541-554.

Yates. A. M. and A. A. Warren. (2000). The phylogeny of the 'higher' temnospondyls (Vertebrata, Choanata) and its implications for the monophyly and origins of the Stereospondyli, *Zoological Journal of the Linnaean Society*, 128, pp. 77- 121.

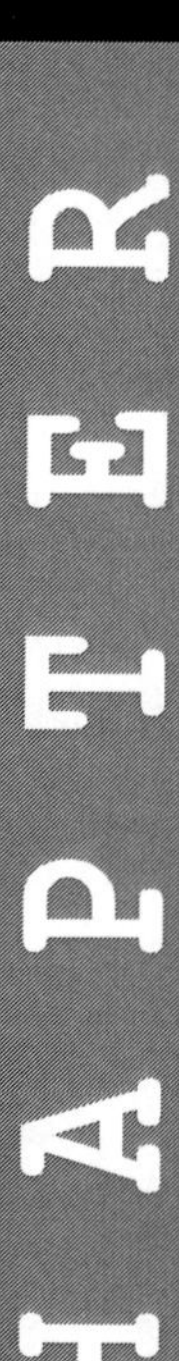

Spatial Mathematical Models for Mineral Potential Mapping

Alok Porwal, E.J.M. Carranza and M. Hale

Abstract

A key problem in spatial-mathematical-model-based mineral potential mapping is the selection of appropriate functions that can effectively approximate the complex relationship between target mineral deposits and recognition criteria. This paper evaluates a series of spatial mathematical models based on different linear and non-linear functions by applying them to base-metal potential mapping of the Aravalli province, western India. Linear models applied are an extended weights-of-evidence model and a hybrid fuzzy weights-of-evidence model, while non-linear models are knowledge and data-driven fuzzy models, a neural network model, a hybrid neuro-fuzzy model and an augmented naive Bayesian classifier model. The parameters of the knowledge-driven fuzzy model are estimated from the expert knowledge, while those of the neural network and Bayesian classifier model are estimated from the data. The two hybrid models use both expert knowledge and data for parameter estimation.

As compared to the linear models, the non-linear models generally perform better in predicting the known base-metal deposits in the study area. Although the linear models do not fit the data as efficiently as the non-linear models, they are easier to implement using basic GIS functionalities and their parameters are more amenable to geoscientific interpretation. In addition, the linear models are less susceptible to the curse of dimensionality as compared to non-linear models, which makes them more suitable for applications to mineral potential mapping of the areas where there is a paucity of training mineral deposits. The hybrid models that conjunctively use both knowledge and data for parameter estimation generally perform better than purely knowledge-driven or purely data-driven models.

6.1 Introduction

Mineral potential mapping of an area involves demarcation of potentially-mineralized zones based on geologic features that exhibit significant spatial association with target mineral deposits. These features, which are termed recognition criteria, are spatial features indicative of various genetic earth processes that acted conjunctively to form the deposits in the area. Recognition criteria are sometimes directly observable; more often, their presence is inferred from their responses in various spatial datasets, which are appropriately processed to enhance and extract the recognition criteria and to obtain evidential or predictor maps.

In traditional approaches to mineral potential mapping, the predictor maps are interpreted, either individually or conjunctively using manual overlay, to demarcate potentially-mineralized zones. In recent years, use of geographic information systems (GIS) for digital overlay of predictive maps has supplanted these traditional approaches. However, naive applications of GIS, which involve, for example, simple overlay or Boolean operations to combine predictor maps, are usually unsuitable for mineral potential mapping because they tend to give equal importance to all recognition criteria. Given the complexity of earth systems that form mineral deposits, it is too naive to assume that all earth processes that were involved in the formation of the target mineral deposits made equal contributions and, hence, all recognition criteria have equal importance as indicators of the target mineral deposits.

It is pragmatic to assume that the process of mineralization responds to random (or stochastic) spatial controls on localization of mineral deposits. Stochastic processes are readily modeled by the application of mathematics (for example, Agterberg, 1974, Alberti and Uhlmann, 1982, Nelson, 1995). It is possible to use appropriate spatial mathematical models of the relation between recognition criteria and target mineral deposits for mineral potential mapping. What is important is the selection of appropriate spatial mathematical model(s).

A spatial mathematical model for mineral prospectivity mapping can be defined as a highly-simplified mathematical representation of the spatial association between exploration criteria (generally represented by predictor maps) and the target mineral deposits. A generalized spatial mathematical model can be empirically represented as below (see also Bonham-Carter, 1994):

$$MPM = f(x_{ij}, P_k)$$

where MPM is a mineral prospectivity map, x_{ij} is the jth (j = 1 to L) pattern on the ith predictor map X_i (i = 1 to N), P_k is the kth (k = 1 to K) parameter of the mathematical function f, L is the total number patterns on X_i, N is the total number of predictor maps and K is the total number of parameters of the function f.

Based on whether the relationship is hypothesized to be linear or non-linear, a variety of linear and non-linear functions can be used to approximate the relationship between recognition criteria and mineral deposits (Bonham-Carter, 1994, Porwal, 2006). The most widely used approaches to spatial-mathematical-model-based approaches to mineral potential mapping include weights-of-evidence (Agterberg *et al.*, 1990, Bonham-Carter Agterberg, 1990), fuzzy approach (An *et al.*, 1991, Getting and Bultman, 1993, Bonham-Carter, 1994, Knox-Robinson, 2000); fuzzy weights-of-evidence (Cheng and Agterberg, 1999) and neural network (Singer and Kouda, 1996, 1997a, 1997b, 1999, 2003) approaches. However, the main problem in spatial-mathematical-model-based approaches to mineral potential mapping is the selection of appropriate functions that can effectively approximate the relation between the target mineral deposits and recognition criteria as well as account for dependencies amongst the recognition criteria. Therefore, a key question is: which mathematical function(s) can be used most effectively to approximate the relationship between a set of recognition criteria (or predictor maps) and the target mineral deposits?

In this review paper, the limitations and strengths of a variety of linear and non-linear spatial mathematical models are examined by applying them to base-metal potential mapping of a study area in the Archean-Proterozoic Aravalli metallogenic province of India. The study area measures about 50,000 km^2 and is located between latitudes 23^{0}30′N and 26^{0}N and longitudes 73^{0}E and 75^{0}E (Fig. 6.1).

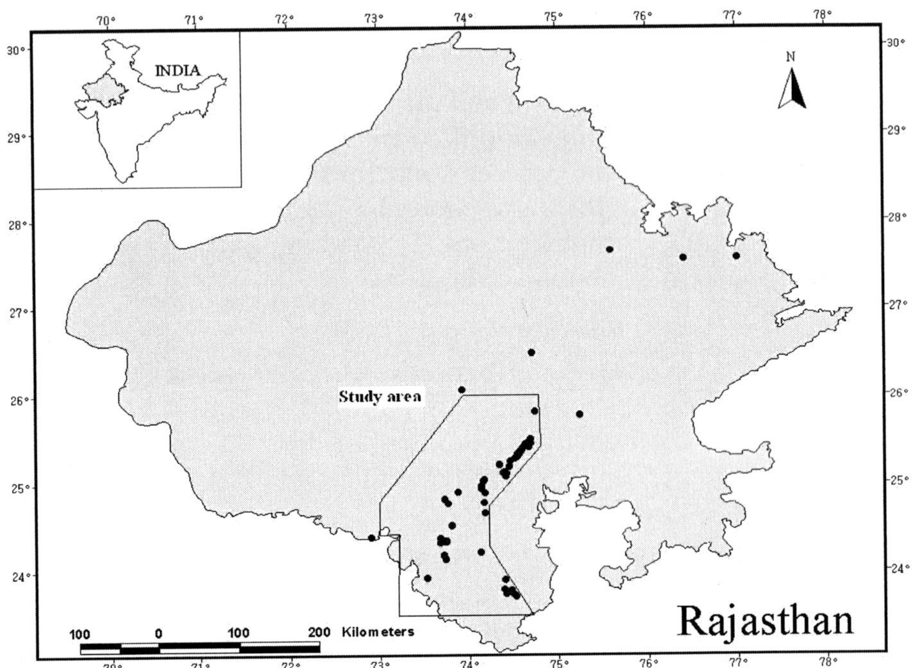

Fig. 6.1 : Location map of study area in the state of Rajasthan, India. Small black circles are the base metal deposits.

6.2 Generalized Methodology

A flow chart of the generalized methodology for implementing spatial-mathematical-model-based mineral potential mapping is shown in Fig. 6.2. The main steps involved in developing and implementing the spatial mathematical models for base-metal potential mapping in the study area are described below.

6.2.1 Step 1: Identification of base-metal Deposit Recognition Criteria

A conceptual approach (Hodgson and Troop, 1988) was used for identifying regional-scale (1:250,000) recognition criteria for base-metal deposits in the study area and representing them as predictor maps for inputting to the spatial mathematical models. A model of base-metal metallogenesis in the framework of overall tectono-stratigraphic evolution of the Aravalli province was developed based on published studies coupled with new interpretive syntheses of regional-scale exploration data (Porwal *et al.*, 2006b). Based on this model, 1) host rock lithology, 2) stratigraphic position, 3) (palaeo-) sedimentary environment, 4) association of synsedimentary mafic volcanic

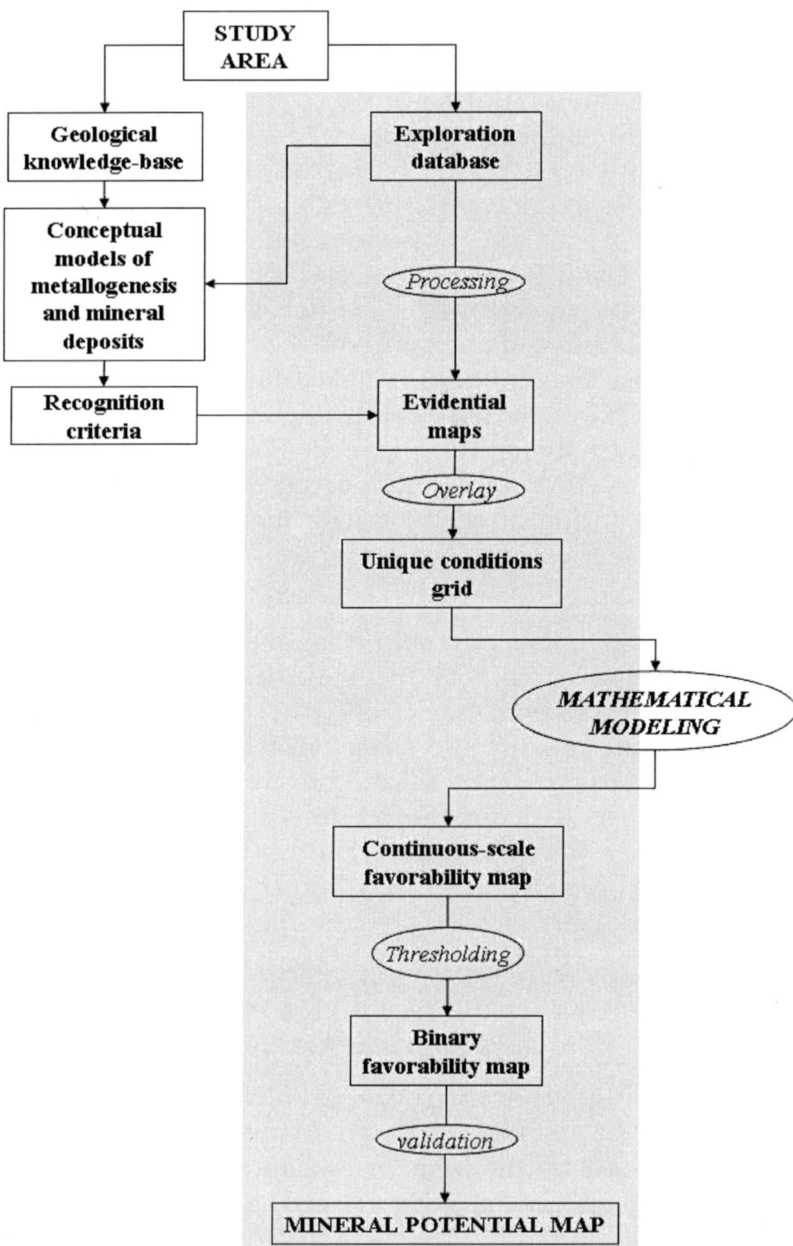

Fig. 6.2 : Flowchart of generalized procedures used for mineral potential mapping. Small rectangular boxes contain objects created at various stages and small elliptical boxes contain processes used for creating the objects. Shaded area represents GIS environment and contains objects and processes that were, respectively, created and implemented within the GIS. Mathematical modeling was implemented, depending on the model, both inside and outside the GIS and so lies partially inside the shaded area.

rocks, 5) proximity to regional tectonic discontinuities and 6) proximity to favorable structures as the most significant regional-scale recognition criteria. The first three criteria are based on the stratigraphically-controlled synsedimentational nature of mineralisation. The fourth criterion is based on the suggestion of Deb (1999) that synsedimentary mafic volcanic rocks in ore environments provided the heat necessary for the circulation of exhalative brines and could be a significant source of metals as well. The fifth criterion is based on our interpretation that the regional tectonic discontinuities, which mark the boundaries of major intracratonic rift sequences (extensional faults) in the province could have served as conduits for heat sources as well as for the hydrothermal discharge as evidenced by their close spatial association with synsedimentary mafic volcanic rocks (Porwal *et al.*, 2006c). The last criterion is based on extensive postgenetic remobilisation and relocation of ores during subsequent deformation, which is especially evident in the Zawar and Pur-Banera zones.

A regional-scale GIS was compiled by digitizing the lithostratigraphic map (Gupta *et al.*, 1995a), the structural map (Gupta *et al.*, 1995b) and the map of total magnetic field intensity (GSI, 1981). The locations of 54 known base metal deposits/occurrences were compiled from various sources. A subset of 30 to 40 deposits, regarded as "discovered" and randomly selected from the known base metal deposits, was used to train the models and to test conditional independence of evidential maps. The remaining deposits were regarded as "undiscovered" and were subsequently used to validate the data-driven model.

The input maps were processed, interpreted and reclassified to create predictor maps for the recognition criteria. In some maps, redundant classes were merged to suppress noise.

The lithostratigraphic map was initially reclassified into two evidential maps, viz., a lithological and a stratigraphic map. The two maps were analysed on the basis of models of metallogenesis, and classes considered redundant were merged.

The sedimentary environment of each stratigraphic formation was interpreted on the basis of published studies (Roy and Paliwal, 1981, Gupta *et al.*, 1997) and available information on mineralogy, texture and structure of constituent (meta)sedimentary rocks (Pettijohn, 1975). These were mapped to generate the evidential map of sedimentary environments.

The total magnetic intensity field data were digitally processed to calculate the 3-D analytical signals of the total magnetic intensity field, which were interpreted to map the mafic igneous rocks in the province. This map was reclassified and classes having no relation to the base metal mineralisation were merged, resulting in the evidential map of mafic igneous rocks.

A shaded relief image of the total field intensity data was used for interpreting regional magnetic lineaments. These lineaments, which mark the regional tectonic discontinuities, were buffered into five zones, each 2 km wide, up to a distance of 10 km. A sixth zone comprising parts of the study area at a distance greater than 10 km from the regional-scale magnetic lineaments was added to generate the evidential map of buffered distances from regional tectonic discontinuities.

The fold axes from the structural map were buffered into ten zones, each 500 m wide, up to a distance of 5 km. Parts of the study area at a distance greater than 5 km from the fold axes were classified as the eleventh zone to generate the evidential map of buffered distances from fold axes.

The trends of magnetic anomalies were mapped from the shaded-relief image of the total field intensity data. A rose diagram of the lineaments thus obtained indicated four preferred orientations, viz., the NE-SW, NW-SE, E-W and N-S. The last two trends showed no significant spatial association with the base metal mineralisation in the province, and so they were filtered out. The NE- and NW-trending lineaments were extracted as separate maps and each set of lineaments was buffered into 5 zones, each 1.5 km wide, up to a distance of 7.5 km. To these was added a sixth zone comprising parts of the study area at a distance greater than 7.5 km from the lineaments. In this way the two predictor maps of buffered distances from NE-trending and NW-trending lineaments were generated. The maps of buffered distances from fold axes, NE-trending lineaments and NW-trending lineaments were used as predictor maps for favorable structures.

Finally, the above multi-class maps of buffered distances from various linear features were converted into binary maps using the method described by Agterberg *et al.* (1990).

To cross-validate the above recognition criteria, which were identified using the conceptual approach, empirical modeling of spatial associations between known base-metal deposits and the recognition criteria (represented as predictor maps) was performed. It was found that all of the above have a strong positive spatial association with the known deposits (Porwal, 2006).

6.2.2 Step 2: Generation of Unique Conditions Grid

A "unique conditions grid map" was generated by combining predictor maps in the GIS using 1 km^2 as the unit cell size. It is defined as "an integer grid formed by the combination of two or more predictor maps, in which the class values represent uniquely-occurring combinations of the classes of the input themes" (Kemp *et al.,* 2001). An attribute table associated with a unique conditions grid map (unique conditions table) contains one record per unique condition class and one field for each predictor map. The unique conditions grid and table are produced by an overlay of the predictor maps in grid format using Spatial Analyst (ArcView3x/ArcGIS9x). The predictor maps were input to the spatial mathematical models in the form of a unique conditions grid.

6.2.3 Step 3: Spatial Mathematical Modeling

A series of spatial mathematical models based on a variety of linear and non-linear function were developed and implemented for base-metal potential mapping in the study area. The linear models applied in this research are an extended weights-of-evidence model (described in Porwal *et al.,* 2003b) and a hybrid fuzzy weights-of-evidence model (described in Porwal *et al.,* 2006c), while the non-linear models are knowledge and data-driven fuzzy models (described in Porwal *et al.,* 2003a), a neural network model (described in Porwal *et al.,* 2003c), a hybrid neuro-fuzzy model (Porwal *et al.,* 2004) and an augmented naive Bayesian classifier model (Porwal *et al.,* 2006a). These models were implemented either inside the GIS or outside the GIS using specialized computer programs, depending on the software resources required to implement their parameter-learning algorithms. Similarly, procedures used for implementing each spatial mathematical model depend on the specific requirements of the model and have been described in the relevant papers cited above. In general, all spatial mathematical models were implemented in the following two steps.

1. Each model was first trained by estimating the model parameters either heuristically from conceptual knowledge or algorithmically from training data.

2. The trained model was then used to process the unique conditions grid. The output for each unique condition was stored in a new field in the unique conditions table.

The parameters of the knowledge-driven fuzzy model were estimated from the expert knowledge, while those of the extended weights-of-evidence, neural network and Bayesian classifier models were estimated from the data. The parameters of the data-driven fuzzy, neuron-fuzzy and fuzzy weights-of-evidence models were estimated using both expert knowledge and the data.

6.2.4 Step 4: Generation of Continuous-scale Favorability Maps

The outputs of each spatial mathematical model for the unique conditions were mapped in the GIS to generate a continuous-scale favorability map (if the modeling was implemented outside the GIS, the processed unique conditions grid was imported back into the GIS). For each unique condition, the output of the spatial mathematical model, irrespective of its form, was interpreted as a relative favorability value. Fig. 6.3 shows the grey-scale base-metal potential maps of the study area derived using the above models.

6.2.5 Step 5: Generation of Binary/ternary Favorability Maps

Continuous-scale favorability maps were cumbersome to interpret for demarcating areas of mineral potential because they represent favorability in a continuous scale from 0 (minimum) to 1 (maximum). Thresholding the favorability values based on some objective criteria facilitates the demarcation of mineral potential zones. In this research, a graphical procedure based on the hypothesis that mineral-bearing areas occupy a very small proportion of the total area of a metallogenic province (Boleneus *et al.*, 2001) was used to identify threshold favorability values. The threshold favorability values were identified based on the inflexion points on favorability versus area curves (see Porwal 2006, p. 119-120). Depending on whether the curve contained one or two inflexion points, one or two threshold favorability values were identified, which were then used to reclassify the continuous-scale favorability maps into binary or ternary favorability maps (Fig. 6.4).

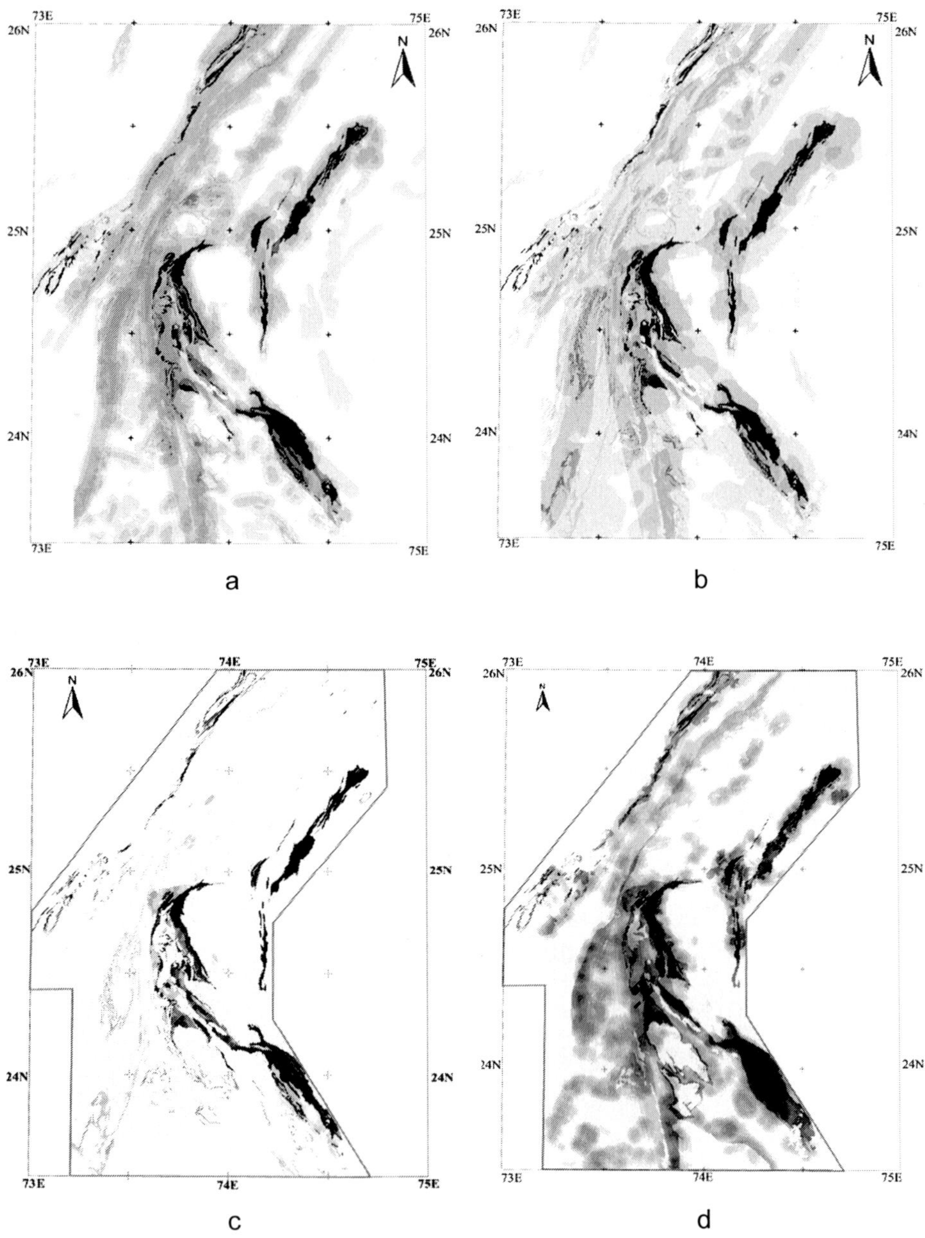

Fig. 6.3 : Continuous scale favorability maps (in inverted grey scale) generated by a) knowledge-driven fuzzy model (favorability ranges from 0.00 (white) to 1.00 (black), b) data-driven fuzzy model (favorability ranges from 0.00 (white) to 1.00 (black), c) extended weights-of-evidence model (favorability ranges from 0.00 (white) to 0.99 (black), d) hybrid fuzzy weights-of-evidence model (favorability ranges from 0.00 (white) to 0.21 (black), *continued.....*

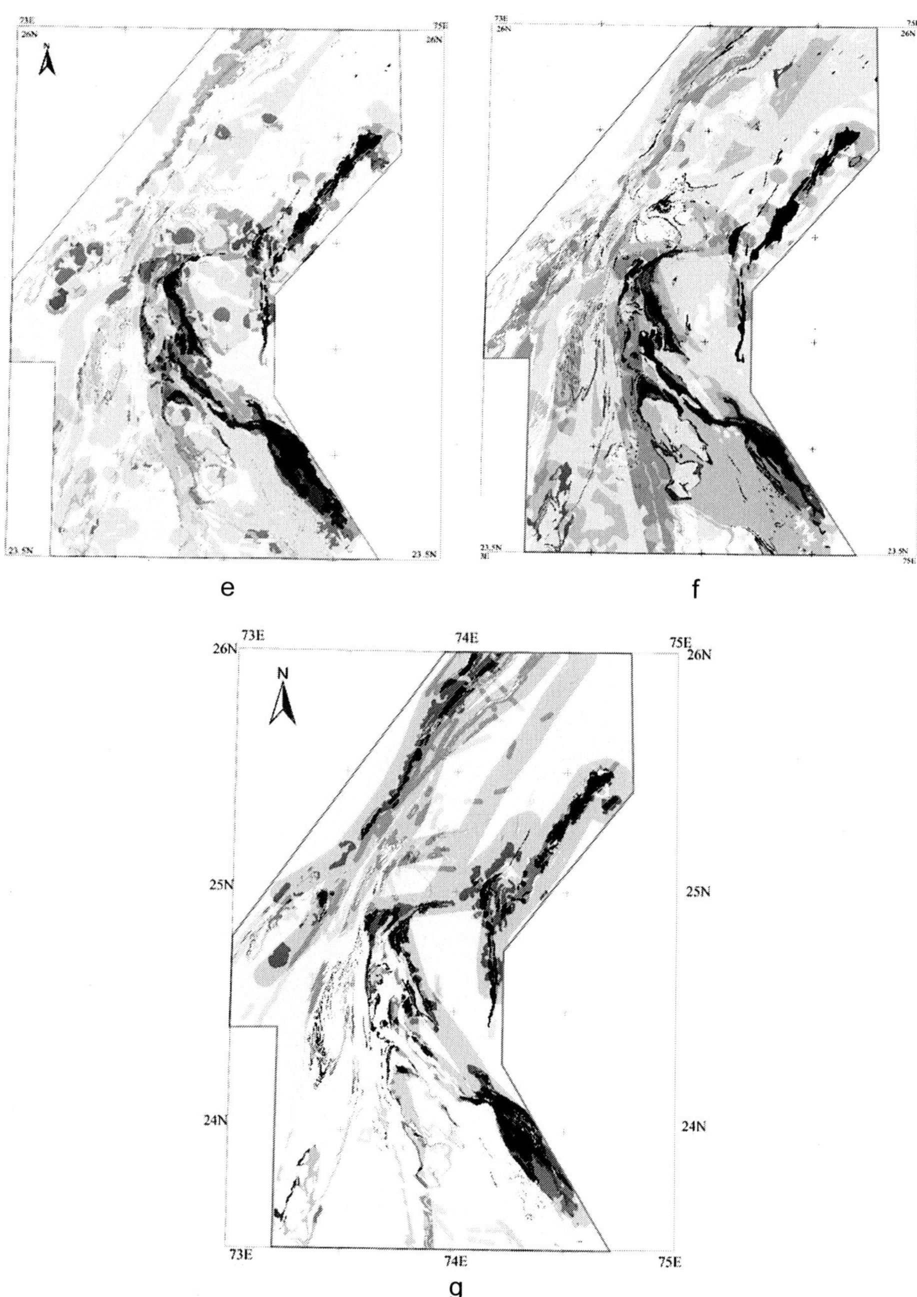

Fig. 6.3 : e) artificial neural network model (favorability ranges from -0.14 (white) to 1.34 (black), f) hybrid neuron-fuzzy model (favorability ranges from 0.00 (white) to 1.00 (black), and g) augmented Bayesian classifier model (favorability ranges from 0.00 (white) to 1.00 (black).

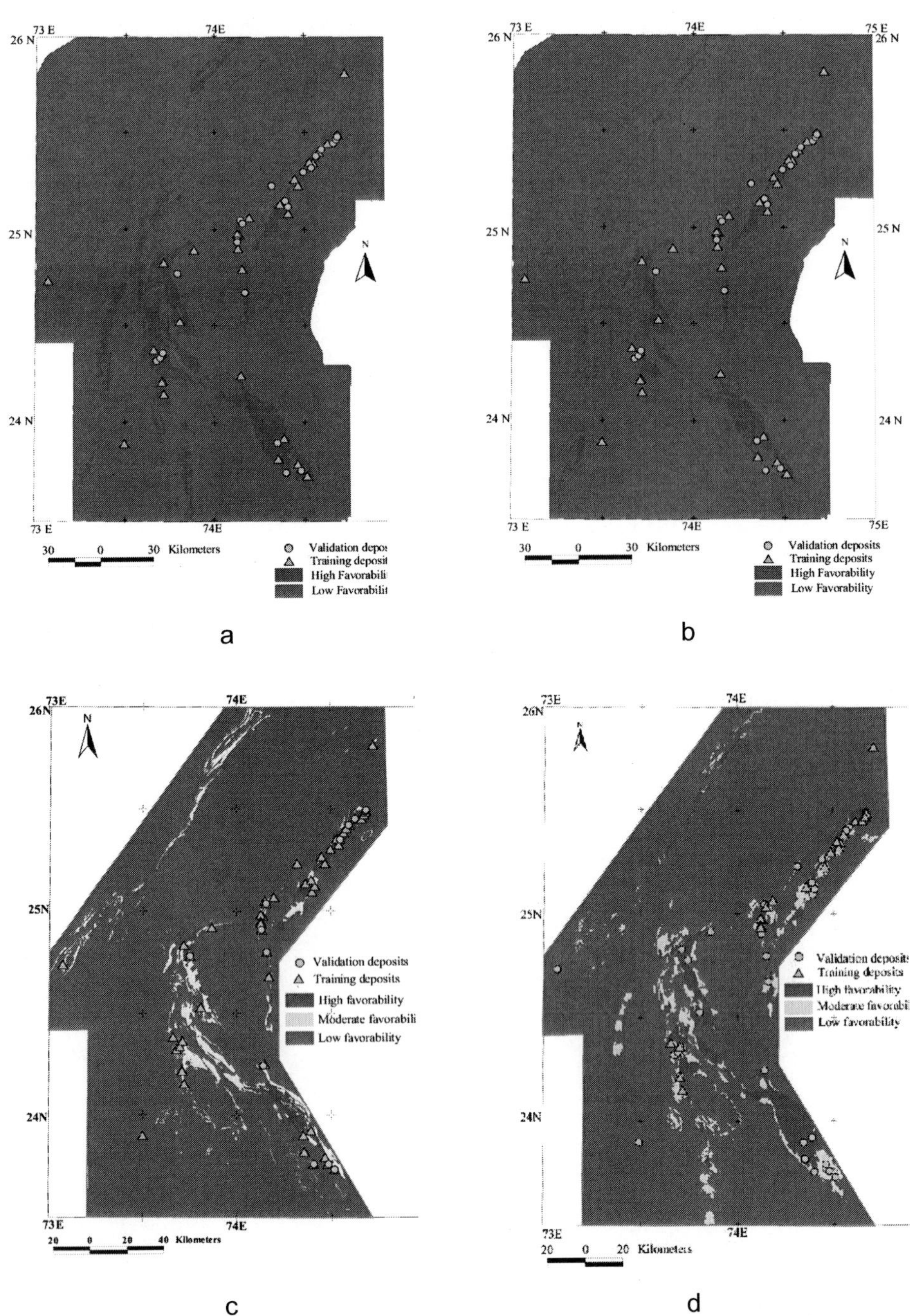

Fig. 6.4 : Binary (or ternary) favorability maps generated by a) knowledge-driven fuzzy model, b) data-driven fuzzy model, c) extended weights-of-evidence model, d) hybrid fuzzy weights-of-evidence model, *continued ....*

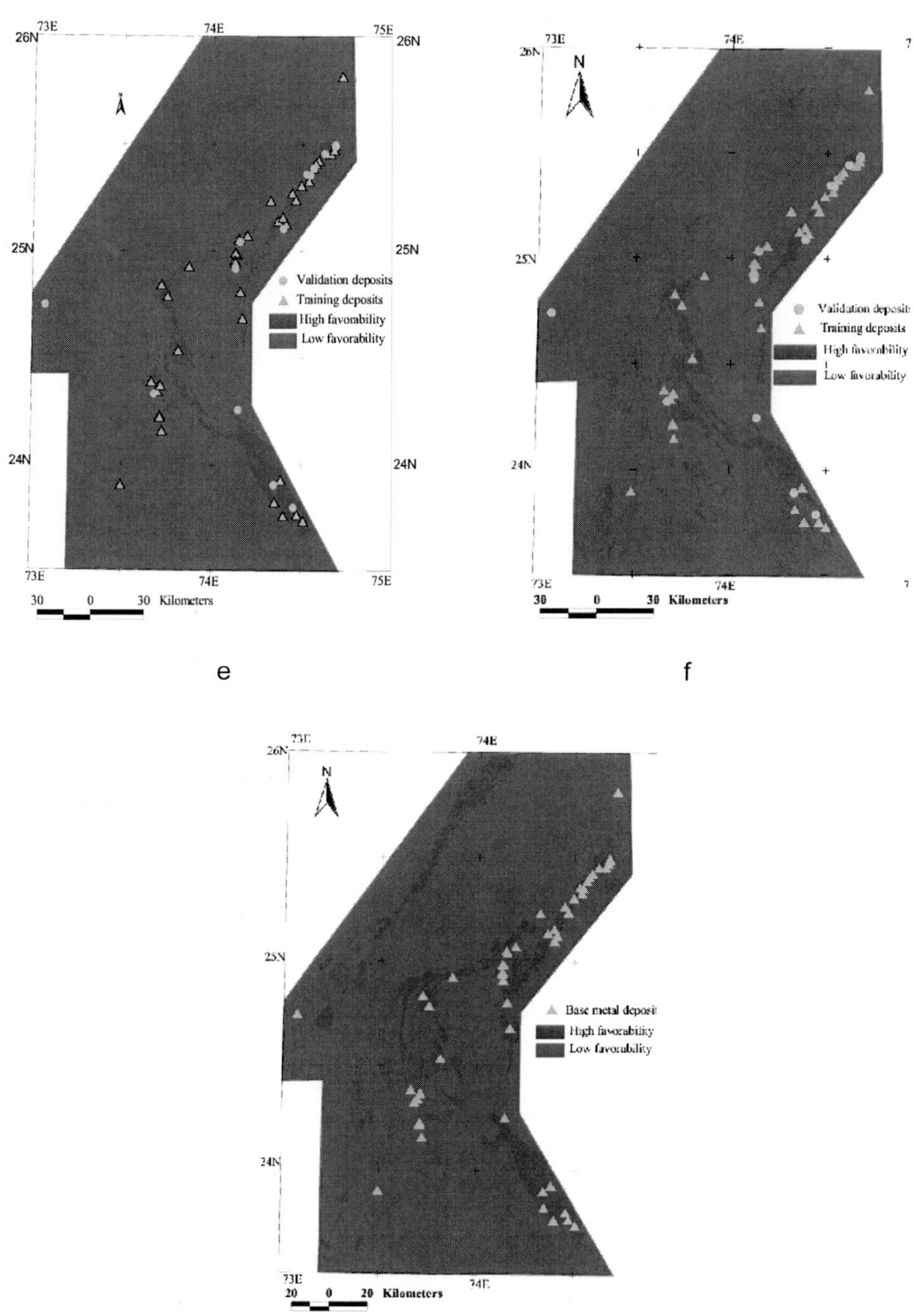

Fig. 6.4 : e) artificial neural network model, f) hybrid neuron-fuzzy model and g) augmented Bayesian classifier model.

6.2.7 Step 6: Validation of Favorability Maps

The binary favorability maps were validated by overlaying locations of known deposits chosen for validation on the binary favorability maps. After the validation, the binary favorability maps could be used as base-metal potential maps of the study area.

6.3 Discussion

An evaluation of the linear models and the non-linear models based only on their performance in classifying training and validation deposits in high favorability zones is not appropriate for three reasons. First, the knowledge-driven fuzzy model did not require separate sets of validation and training deposits and, hence, all deposits were used to validate it. Second, a different set of validation and training deposits was used in the data-driven fuzzy model and the fuzzy weights-of-evidence model in order to test the ability of these hybrid models to function with smaller numbers of training points. Third, an n-fold cross validation method was used for validating the augmented Bayesian classifier model, while all other models were validated using a hold-back validation method.

The models are evaluated based on a number of criteria, which include percentage of correctly-classified known deposits, theoretical requirement of conditional independence, interpretability of model parameters, ease of implementation, susceptibility to the curse of dimensionality and ability to quantify uncertainties (Table 6.1). The ability of a model to conjunctively use both conceptual and empirical components of available geoscientific information of an area is also used as an evaluation criterion, because it allows the model to compensate the deficiencies of one component with the other component. A correct classification of the Rampura-Agucha deposit is also included as an evaluation criterion. Rampura-Agucha, the only world-class Zn-Pb deposit of India, is a difficult deposit to predict, not least because of its unusual regional geological setting (Porwal et al., 2003a). The correct classification of the deposit can therefore be interpreted as an indication of the ability of a spatial mathematical model to predict mineral deposits in unusual geological settings. Table 6.1 summarizes the performance of the linear and the non-linear spatial mathematical models with respect to the above evaluation criteria.

Table 6.1 indicates that both linear and non-linear models demarcate high favorability zones that occupy less than 10% of the search area and predict more than 80% of the known base-metal deposits. This indicates that both categories of models can efficiently narrow down search areas for mineral exploration. However, the non-linear models generally perform better in predicting the known base-metal deposits (Table 6.1). This suggests that, given the complexity of earth systems that result in the formation of mineral deposits, non-linear functions more adequately approximate the relation between predictor patterns and target mineral deposits and therefore fit the data more efficiently. This interpretation is reinforced by a comparison of the predictive classification results of the linear and the non-linear models with respect to the Rampura-Agucha deposit. The deposit is incorrectly classified in low favorability zones by all linear models, whereas two of the non-linear models, namely, the neural network model and the hybrid neuro-fuzzy model, correctly classify the deposit. The correct classification of the Rampura-Agucha deposit by two of the non-linear models may be attributed to the ability of the non-linear functions used in the two models to recognize the critical predictor pattern(s) and respond in a highly non-linear way to maximize the contribution of such component(s) to the output.

The linear models, which are based on simplified and modified Bayesian equations, require conditional independence of input predictor maps (Table 6.1) and by assuming it, they ignore the effects of possible interactions amongst input predictor patterns. The assumption is often difficult to validate, although several statistical tests have been proposed by different workers to test conditional independence (Singer and Kouda, 1999, Porwal et al., 2003b). Given the peculiar nature and interdependency of geological processes that result in mineral deposits, the possibility of conditional dependencies amongst predictor patterns is often hard to rule out. The moot question, however is: how seriously is the performance of these Bayesian models affected by conditional dependencies in input data? The question was explored by Porwal et al. (2006a) and there appears to be some evidence that their performance is not seriously degraded, especially if the output posterior probabilities are interpreted as relative favorability values. Nevertheless, as shown by Porwal et al. (2006a), the performance of a Bayesian model improves if it is implemented in a non-linear mode without assuming conditional independence of input predictor patterns. The better performance of the non-linear models as compared to the linear models may therefore be partially attributed to their ability to recognize and compensate for conditional dependencies amongst input predictor patterns.

Table 6.1 : Performance of spatial mathematical models with respect to some evaluation criteria.

Evaluation criterion	Linear		Non-linear				
	Weight-of-evidence model	Hybrid fuzzy weight-of-evidence model	Knowledge-driven fuzzy model	Data-driven fuzzy model	Neural network model	Hybrid neuro-fuzzy model	Augmented naïve Bayesian classifier model
% of known deposits classified in high favorability zones	83.3	87	87	85.2	94.4	96.3	88.9
% of study area classified in high favorability zones	3.1	5.9	8.9	5.4	6	9.5	11.3
Whether Rampura-Agucha deposit classified in high favorability zones	No	No	No	No	Yes	Yes	No
Assumption of conditional independence	Required	Required	Not required	Not required	Not required	Not required	Not required
Whether parameters easily interpretable	Yes	Yes	No	No	No	No	Yes
Whether parameter learning easily implemented in GIS	Yes	Yes	Yes	Yes	No	No	No
Whether affected by the curse of dimensionality	Not significantly	No	No	Not significantly	Yes	Yes	Yes
Whether uncertainty quantified	Stochastic uncertainties quantified	Yes	No	No	No	No	No
Basis of parameter estimation	Data	Both data and Knowledge	knowledge	Both data and knowledge	Data	Both data and knowledge	Data

However, although the simplified and modified Bayesian equations used in the linear models may not fit the data perfectly, they allow significant insights into the data. The parameters of the linear models can be interpreted easily (Table 6.1) to gain insights into the relative contributions of various predictor patterns (or recognition criteria) and, hence, into genetic processes that formed target mineral deposits (Bonham-Carter, 1994). The non-linear functions, on the other hand, may fit the data more efficiently, but their parameters are generally not amenable to direct geoscientific interpretations.

The linear models also have an advantage over the non-linear models in that they are easily implemented in a GIS environment (Table 6.1). The non-linear models, on the other hand, require specialized computer programs for parameter estimations and, therefore, are generally implemented outside the GIS and the output values are imported back into the GIS in order to generate favorability maps. (The fuzzy models are exceptions in this respect, but that is because their parameters are estimated heuristically from knowledge and, therefore, do not require algorithmic estimations.)

The linear models are, in general, not seriously affected by high dimensionality of input data (Table 6.1). On the other hand, the non-linear models (again, with the exceptions of the fuzzy models) are more susceptible to the curse of dimensionality and, if implemented without caution, may over-fit the training data. Although over-fitting can be controlled by following efficient training procedures (Porwal et al., 2003c), it is often necessary to reduce dimensionality of the input data in order to avoid it.

Quantification of uncertainty in the output of a model is important because it facilitates effective decision making. The linear models used in this research quantify uncertainties in their outputs. The non-linear models, on the other hand, do not quantify the uncertainties in their outputs.

Most spatial mathematical models, both linear and non-linear, are implemented either in a data-driven mode or in a knowledge-driven mode, which entails that a significant proportion of the geoscientific information remains under-utilized. Moreover, these models cannot 'cross-compensate' possible deficiencies in either the knowledge-base or in the database. This research attempted to address this problem by applying 'hybrid' models that can conjunctively use both data and knowledge for parameter estimation. Table 6.1 shows that, in both linear and non-linear categories, the hybrid models perform better than the purely knowledge-driven or purely data-driven models.

6.4 Conclusions

Based on the above discussions, the following conclusions can be drawn.

1. Both linear and non-linear models can be effectively used for narrowing down search areas for mineral exploration.
2. The non-linear models perform better, as compared to the linear models, in predicting base-metal deposits in the study area, which indicates that, as compared to the linear functions, the non-linear functions more adequately approximate the relationship between predictor patterns and target mineral deposits.
3. The non-linear models, especially the neural network and the hybrid neuro-fuzzy models, can perform better than the linear models in predicting deposits with unusual geological characteristics.
4. The non-linear functions have the ability to take into account the possible conditional dependencies amongst predictor pattern and therefore the response of the non-linear models is conditioned in such a way that the contribution of conditionally-dependent patterns is minimized.
5. The parameters of the linear models are amenable to direct geoscientific interpretations. On the other hand, the parameters of the non-linear models can not always be interpreted to gain insights into the data and genetic processes.
6. The linear models are conveniently implemented in a GIS-environment, where as the non-linear models require specialized computer software for implementation.
7. A susceptibility to the curse of dimensionality is a major limitation of the non-linear models.
8. Hybrid spatial mathematical models perform better than purely knowledge-driven and purely data-driven models.

Amongst the spatial mathematical models discussed above, there is no 'best' model that can be effectively used for mineral potential mapping in all situations. The selection of the best model depends mainly on a) geoscientific information available for the modeling, b) computer hardware and software resources available for the modeling and c) whether gaining insights into the data is also an objective of the modeling. The above conclusions provide rough guidelines for selecting the best model in a given modeling situation.

Acknowledgements

This review paper is based on the doctoral research work carried out by the first author at the International Institute for Geo-information Science and Earth Observation (ITC), Enschede, the Netherlands, on a Netherlands Fellowship Programmes (NFP) fellowship. Insightful comments of Profs. Greame Bonham-Carter, Don Singer, Gary Raines, Richard McCammon, Mihir Deb, Alfred Stein, and Richard Howarth on various parts of the research work are gratefully acknowledged.

Bibliography

Agterberg, F.P. (1974). Automated contouring of geological maps to detect target areas for mineral exploration, *Mathematical Geology*, 6, pp. 373-395.

Agterberg, F.P., Bonham-Carter, G.F. and Wright, D.F. (1990). Statistical pattern integration for mineral exploration, in G. Gaal and D.F. Merriam (eds.), Computer Applications in Resource Estimation Prediction and Assessment for Metals and Petroleum. Pergamon Press, Oxford-New York, pp. 1-21.

An, P., Moon, W.M., and Rencz, A. (1991). Application of fuzzy set theory for integration of geological, geophysical and remote sensing data, *Canadian Journal of Exploration Geophysics*, 27, pp. 1-11.

Alberti, P.M., and Uhlmann, A. (1982). Stochasticity and Partial Order: Double Stochastic Map and Unitary Mixing, M. Hazewinkel (ed.) Mathematics and Its Applications, 9, Springer, Boston, 132p.

Boleneus, D.E., Raines, G.L., Causey, J.D., Bookstrom, A.A., Frost, T.P., and Hyndman, P.C. (2001). Assessment method for epithermal gold deposits in northeast Washington State using weights-of-evidence GIS modeling, *USGS Open-File Report*, 01-501, 52 pp.

Bonham-Carter, G.F. (1994). Geographic Information Systems for Geoscientists: Modeling with GIS, Pergamon Press, Ontario, Canada, 398 p.

Bonham-Carter, G.F., and Agterberg, F.P. (1990). Application of a microcomputer based geographic information system to mineral-potential mapping, in J.T. Hanley and D.F. Merriam (eds.), Microcomputer-based Applications in Geology, II, Petroleum, Pergamon Press, New York, pp. 49-74.

Cheng, Q., and Agterberg, F.P. (1999). Fuzzy weights of evidence and its application in mineral potential mapping, *Natural Resources Research*, 8(1), pp. 27-35.

Deb, M. (1999). Metallic mineral deposits of Rajasthan, in P. Kataria (ed.), Proceedings of Seminar on Geology of Rajasthan - Status and Perspective, M.L. Sukhadia University, Udaipur, India, pp. 213-237.

Getting, M.E., and Bultman, M.W. (1993). Quantifying favorableness for occurrence of a mineral deposit type - an example from Arizona, *USGS Open-File Report*, 93-392, 23 pp.

GSI (1981). Total intensity aeromagnetic map and map showing the magnetic zones of the Aravalli region, southern Rajasthan and northwestern Gujarat, India (1:250,000), Geological Survey of India Press, Hyderabad, India, 4 sheets.

Gupta, S.N., Arora, Y.K., Mathur, R.K., Iqballuddin, Prasad, B., Sahai, T.N., and Sharma, S.B. (1997). The Precambrian geology of the Aravalli Region, Memoirs of Geological Survey of India, 123, Geological Survey of India press, Hyderabad, India, 262 pp.

Gupta, S.N., Arora, Y.K., Mathur, R.K., Iqballuddin, Prasad, B., Sahai, T.N., and Sharma, S.B. (1995a). Lithostratigraphic map of Aravalli region, 2nd edition (1:250,000), Geological Survey of India Press, Calcutta, India, 4 sheets.

Gupta, S.N., Arora, Y.K., Mathur, R.K., Iqballuddin, Prasad, B., Sahai, T.N., and Sharma, S.B. (1995b). Structural map of the Precambrian of Aravalli region, 2nd edition (1:250,000). Geological Survey of India Press, Calcutta, India, 4 sheets.

Hodgson, C.J., and Troop, D. G. (1988). A new computer-aided methodology for area selection in gold exploration: A case study from the Abitibi greenstone belt, *Economic Geology*, 83, pp. 952-977.

Kemp, L.D., Bonham-Carter, G.F., Raines, G.L., and Looney, C.G. (2001). Arc-SDM: ArcView extension for spatial data modeling using weights of evidence, logistic regression, fuzzy logic and neural network analysis, http://ntserv.gis.nrcan.gc.ca/sdm/.

Knox-Robinson, C.M. (2000). Vectorial fuzzy logic: a novel technique for enhanced mineral prospectivity mapping with reference to the orogenic gold mineralization potential of the Kalgoorlie Terrane, Western Australia, *Australian Journal of Earth Sciences*, 47, pp. 929-942.

Porwal, A. (2006), Mineral Potential Mapping with Mathematical Geological Models, ITC PhD Dissertation, University of Utrecht, ISBN no. 90-6164-240-X, 289 pp.

Porwal, A., Carranza, E.J.M., and Hale, M. (2003a). Knowledge-driven and data-driven fuzzy models for predictive mineral potential mapping, *Natural Resources Research*, 12, pp. 1-25.

Porwal, A., Carranza, E.J.M., and Hale, M. (2003b). Extended weights-of-evidence modelling for predictive mapping of base-metal deposit potential in Aravalli province, western India, *Exploration and Mining Geology*, 10, pp. 155-163.

Porwal, A., Carranza, E.J.M., and Hale, M. (2003c). Artificial neural networks for mineral potential mapping: A case study from Aravalli province, western India, *Natural Resources Research*, 12, pp. 155-177.

Porwal, A., Carranza, E.J.M., and Hale, M. (2004). A hybrid neuro-fuzzy model for mineral potential mapping, *Mathematical Geology*, 36, pp. 803-826.

Porwal, A., Carranza, E.J.M., and Hale, M. (2006a). Bayesian network classifiers for mineral potential mapping, *Computers and Geosciences*, 32, pp. 1-16.

Porwal, A., Carranza, E.J.M., and Hale, M. (2006b). Tectonostratigraphy and base-metal mineralization controls, Aravalli province (western India): new interpretations from geophysical data, *Ore Geology Reviews*, 29, pp. 287-306.

Porwal, A., Carranza, E.J.M., and Hale, M. (2006c). A Hybrid fuzzy weights-of-evidence model for mineral potential mapping, *Natural Resources Research*, 15, pp. 1-15.

Nelson, R. (1995). Probability, Stochastic Processes, and Queueing Theory: The Mathematics of Computer Performance Modeling, Springer-Verlag, New York, 583 p.

Pettijohn, F.J. (1975). Sedimentary Rocks, Harper and Row, New York, 628 p.

Roy, A.B., and Paliwal, B.S. (1981). Evolution of lower Proterozoic epicontinental deposits: stromatolite bearing Aravalli rocks of Udaipur, Rajasthan, India, *Precambrian Research*, 14, pp. 49-74.

Singer, D.A., and Kouda, R. (1996). Application of a feedforward neural network in the search for Kuroko Deposits in the Hokuroku district, Japan, *Mathematical Geology*, 28, pp. 1017-1023.

Singer, D.A., and Kouda, R. (1997a). Classification of mineral deposits into types using mineralogy with a probabilistic neural network, *Nonrenewable Resources*, 6, pp. 69-81.

Singer, D.A., and Kouda, R. (1997b). Use of a neural network to integrate geoscience information in the classification of mineral deposits and occurrences, in A.G. Gubins (ed.), Proceedings of Exploration 97: Fourth Decennial International Conference on Mineral Exploration, pp. 127-134.

Singer, D.A., and Kouda, R. (1999). A comparison of the weights of evidence method and probabilistic neural networks, *Natural Resources Research*, 8, pp. 287-298.

Singer, D.A., and Kouda, R. (2003). Typing Mineral Deposits Using Their Grades and Tonnages in an Artificial Neural Network, *Natural Resources Research*, 12, pp. 201-208.

CHAPTER 7

Enterprise GIS in Geological Survey of India

Asit Saha, Basab Mukhopadhyay and Auditeya Bhattacharyya

Abstract

As part of the Enterprise Information Portal (EIP) project, an Enterprise-wide Geographic Information System has been implemented through which Intranet / Internet users enjoying different access levels can quickly search for, locate and visualize geoscientific information generated by Geological Survey of India (GSI). The proposed system will provide real-time response to information needs and will use secure Internet technology to deliver information to bonafide users. It consists of an application that will enable data owners to load new data sets, update/replace existing data sets in a highly secure environment, load and maintain metadata about their data sets in a Data Catalog and a subsystem that will query, search, locate and display geo-information stored in the centralized Geodatabase.

Enterprise GIS has been deployed primarily using a relational database, a spatial database engine tool, an Internet Map server tool, and integrating the existing ArcGIS desktop GIS clients. The approach is to import data from distributed data sources into the centralized repository using a customized extraction-transformation-loading (ETL) tool, thereby making appropriate structural changes to ensure homogeneity at various levels, and load into the main Geodatabase.

While delineating the functionalities of enterprise GIS solution an earnest attempt was made to accommodate the diverse needs of the organization, including its processes, activities, products and the whole gamut of users at different levels from the field geologist to the decision makers. An effort was also made to standardize various data automation processes and to iron out the heterogeneities existing in legacy data. Ultimately, GSI aspires to create a seamless geospatial database for the entire geographic extent of the country, which can be easily shared by the geoscientific community.

7.1 Introduction

Geoscientists now face the challenge of managing the multidisciplinary and complex information in a cost effective way. The information provided by geological surveys are used by a variety of stakeholders from public, private, industrial and academic sectors and is important for the users' respective decision making tasks. This necessitates an integrated infrastructure to facilitate data sharing and workflow management, knowledge integration, decision support and dissemination.

Geological Survey of India was established in 1851. The first charter of GSI was to cover the country by systematic geological survey. Over the years, GSI has taken up the responsibility of integrated natural resource exploration, natural hazard mitigation, studies in Antarctica, maintenance and dissemination of geoscientific information, etc (URL 1).

The voluminous geospatial data collected by GSI over more than 150 years are presently available as field reports, maps and different other publications in distributed centres all over the country. Data exists in both analog and digital form and storage, compilation, dissemination of the same is manually managed. This has resulted in problems like heterogeneous, duplicate data, lack of documentation about their usability, availability and completeness and very limited accessibility.

To tackle these problems, GSI has developed a fully integrated Enterprise GIS (EGIS), replete with high quality, easily available data, and a robust tool to view, explore and analyze the data (URL 2). This paper discusses the rationale behind that initiative from GSI's perspective and describes the design / architecture that the EGIS adopted and its relationship with the organizational requirements and constraints.

The core users of the system are all officers associated with field projects, Map & Cartography, Geodata divisions, Planning & Monitoring divisions, etc. The EGIS will also benefit users who do not use a sophisticated GIS tool, and will enable field geologists, to get better access to the data they require for their work. It will cater to the need for knowledge management, improved collaboration and communication amongst the broader community and to keep pace with the national level initiatives like e-governance (URL 3) and the National Spatial Data Infrastructure movement (URL 4). This EGIS initiative is a part of a bigger Information Infrastructure of GSI, which comprises the GSI Intranet and GSI Enterprise Information Portal (EIP) (Saha and Ganguly, 2003).

7.2 Background Information

7.2.1 Organizational Structure

It is important to have a comprehensive idea about the organizational structure of GSI (GSI, 2006), inasmuch as it has a direct bearing on the workflow design of Enterprise GIS. The GSI functions under the Ministry of Mines (MOM) and is headed by the Director General. The overall responsibility of planning, programming, monitoring of scientific activities, dissemination of information and interaction with the user community rests with the Director General and his team of senior managers. There are six Regions (geographically based), three specialised Wings (activity based), Training Institute, and the Central Headquarters (CHQ). Besides having functional units in the respective headquarters (RHQ), each of the Regions comprises State based Operational Units (OU). The specialised Wings also comprise sector wise functional units in addition to the headquarters setup. Normally, SAG level officers co-ordinate the activities related to the programme formulation, administration and technical control in all these units.

The primary function of collecting basic geological information is carried out through the Divisions/Projects spread all over the country, which are functional entities with specific work schedule, time frame, manpower and material inputs. Clusters of such Divisions and/or Projects, normally headed by JAG level officers are located at the Regional and Operational offices and in many outlying stations, totaling

33 cities/towns in the country. These JAG level officers initiate actions for programme formulation, coordinate and supervise the approved project work, interact with other disciplines and finalise the scientific and technical reports. Divisions/Projects are provided with a number of JTS/STS level officers of one or more streams, who actually constitute the field parties engaged in data collection, its analysis, synthesis and dissemination.

The support activities to the geological investigations are provided by independent setups, in complementary disciplines like geophysics, chemistry, drilling, mechanical engineering, materials management, finance and administration. Each of these establishments is based at Regional or Operational offices and is controlled by JAG / SAG level officers. This structure ensures availability of technical guidance and supervision from different disciplines. Periodical review mechanisms at different management levels ensure effective monitoring and programme implementation.

7.2.2 Overview of GSI Data

The essential condition for establishing a geological map is a precise field survey, based on numerous measurements, observations and inferences. In GSI, baseline field survey is normally done at a scale of 1:50,000 or 1:25,000. Besides the outcrops, data from bore-holes, trenches or pits are also taken into consideration. In most cases, complementary investigations are carried out in the laboratory. Apart from geological studies, GSI is also engaged in geochemical, geophysical and various other multidisciplinary geoscientific investigations. Raw data thus collected exists in working mapsheets, field notebooks, field and laboratory datasheets, etc. Technical reports are prepared through synthesis and compilation of the raw data primarily for in-house consumption. GSI also publishes various thematic maps, books, journals and periodicals for general use. Traditionally, map drafting and compilation has been done manually using Mylar sheets or plastic-coated non-expandable papers (Saha and Ganguly, 2003). Nowadays CAD and GIS are increasingly being used for digitization, edge-matching, and all other mapmaking activities. Till late nineties paper maps and technical reports were the dominant form of 'database' in GSI.

7.2.3 GIS Activities in GSI

During the nineties, isolated GIS activities started in different centres of GSI. These consisted of project-level systems with limited ad-hoc sharing of data. Geoscientists were engaged in a broad range of activities having specific goals. They generally utilized baseline geoscientific data and derived new information through modeling and analysis. There was inherent heterogeneity in objective and approach of each project and the resultant datasets used to have a relatively short shelf life. Once the analysis results were published, they were either forgotten or abandoned for the sake of new projects. More than often these project-level geospatial data were not documented or captured. A systematic approach was thus needed to archive, collate and integrate the valuable information gathered by Geological Survey of India through years. GSI took a giant leap forward towards this direction through the implementation of GSI-BRGM Project 'Setting up of Geoscientific Data Centre at GSI' which equipped GSI with a state-of-the-art facility and trained human resource (GSI, 1997). The project had many components, the most significant being the building up of National Geoscientific Database (NGDB) using Oracle Relational database (RDBMS). This was the first organizational scale effort in geospatial data management. This effort focused on building information management infrastructure, which includes a central geospatial data repository, along with standards, policies and guidelines. This was done to ensure data quality and streamlined workflow processes. The project resulted in establishing six Regional centres equipped with state-of-art infrastructure and trained human resources. Next step was to scale up this effort throughout all the operational units of GSI. With this goal in sight Project Geoinformatics was launched in 1999.

7.2.4 Project Geoinformatics

Project Geoinformatics was an organisation-wide effort and it was designed after taking into account the spatial data generation workflow in GSI. Through years the field geologists from various divisions / projects have been collecting data encompassing various themes. The data and the derived information pass through certain quality control and are published as GSI field reports. Within the scope of this project geoscientific information acquired / generated all over the country got collated and integrated in NGDB schema. NGDB consisted of thematic domains like Map50K (baseline geological data collected at scale 1:50,000), Coal (Coal exploration data), Drilling (subsurface mineral investigation data), Geochemical (geochemical exploration data), Geophysical (airborne and ground geophysical exploration data), etc. GSI collects base information using SOI 1:50,000 scale topographic sheets, so that was used as the base unit for compilation. All existing reports pertaining to one sheet was studied, transcribed and the information was entered through forms in respective domains into the NGDB. Corresponding spatial data was automated in ArcINFO and stored separately as coverages, a file-based system. The system worked in client-server architecture, with clients having ArcINFO to automate the geometry and Oracle forms to enter data. A mandatory attribute "Geometry-ID" existed in both the attribute tables of coverages as well as in the NGDB to relate the geometry with the attribute data in NGDB schema. The Geometry-ID was an effort to store a numeric code based on a standardized classification scheme (Croswell, 2000). For example, for the Lithology layer in Map50K domain, it was supposed to contain a code representing the stratigraphic position of the mapunits based on the unified stratigraphic legend of India. Unit offices all over the country replicated this infrastructure and started collating information in a tile-based manner. The system adopted distributed data storage, but following enterprise standards. In this system datasets remained with the data owners, in this case the Regional / Operational units. Each project also sent periodic updates to the Central Repository in the form of Oracle exports and ArcINFO interchange files, which were imported and collated manually to maintain currency.

7.2.5 Problems Faced

As evident from the above discussion, the attribute 'Geometry-ID' was most necessary to establish a relation between the geometry and the attributes. Geometry-ID was supposed to be based on a unified stratigraphic legend (Saha *et al.*, 2003), but considering the complexity of Indian Geology, it was not possible at the beginning to adopt a unified legend - consequently, different regional offices adopted legends constructed on the basis of regional geology. This introduced heterogeneity. A map-unit having same stratigraphic position extending over different Regions continued to have different codes, thus making it almost impossible to correlate and prepare a seamless, integrated map of India.

Spatial data was stored in file-based data models such as coverages (URL 5). There are several significant limitations to any file-based data model. Concurrent user access typically degrades performance and it is not possible to support multiple concurrent users editing a single file. Though organization-wide data standards such as a mandatory attribute 'Geometry-ID', standardized layer nomenclature, standardized layer classification were in place to ensure certain data homogeneity, there was no mechanism in this data model to ensure and verify adoption of these standards. Various projection systems were used across Regions; sometimes files even remained without georeferencing, in raw table coordinates. To summarise, spatial files were generated in an isolated way in different locations, breeding heterogeneity. Lot of manual efforts were needed to harmonise data from different Regions and at times it was often found that the geometry-ID values in coverages and in NGDB do not correspond to each other. There were other technical and economic problems. Every time one wished to view and query the data, connection had to be established manually to the Oracle database through ArcINFO. There was no coherent, easily accessible interface for exploratory visualization. To have a mere glimpse, one has to depend upon a set of key individuals, who use data daily and was generally well informed about the location, content and currency of the geospatial data. A more automated, efficient method of data access was thus needed to alleviate the pressure on those individuals as well as for better organizational knowledge management. There was also the burgeoning cost of maintaining similar infrastructure at distributed centers.

7.3 Needs and Requirements

An Enterprise GIS design is aimed towards efficient sharing of standardized spatial data and to achieve that goal it is important to understand the organizational needs, constraints and roles and needs of each stakeholder.

7.3.1 Organizational Perspective

The Geoinformatics project attempted for the first time to build a central repository of standardized geospatial data. But overall data sharing was inefficient due to the lack of infrastructure and thus the possibility of redundancy and duplicated efforts remained unattended. Project Geoinformatics introduced the enterprise policies for data format, but it was up to the individual users to follow the standards and that proved to be ineffective on an enterprise scale. Since the spatial data existed in file systems, it was difficult to be checked for adherence to standards. There was also no robust mechanism to integrate, manage, share and visualize the data. The problem was less grave for the attribute data stored in NGDB, since the database through its multitude of integrity constraints ensured quality and adherence to standards. Simply put, it was necessary to provide secure, dependable and concurrent access to centrally managed geospatial data to the users. This basic requirement also necessitated the auxiliary requirements of security, record level locking, edit conflict resolution, etc.

7.3.2 Security Perspective

While making a centralized repository accessible to all users, the issues to be taken into consideration are the editing rights and needs of the users, layers which should be visible to which groups of users, sensitivity of the data, etc. The repository would be largely an Intranet resource, so it should be protected by appropriate network infrastructure and firewalls. The intention was to provide authorized access to a definite set of people.

7.3.3 User Perspective

Witkowski *et al.* (2002) highlighted the importance of recognizing the role of the participants while developing a conceptual design for implementation of Enterprise GIS. They identified three categories of participants or stakeholders, *data providers,* data *managers and GIS users.* An assessment of the existing spatial data and information needs of the organization was conducted and the diverse needs of the different groups of users (Witkowski *et al.,* 2002) in GSI were brought out. In GSI, the users belong to three categories: Data owner / creator, data managers, and the data users. Specific users may play a single role or multiple roles in Enterprise GIS. Table 7.1 lists the roles and requirements of the different users from GSI perspective.

To facilitate organization-wide sharing of standardized, current and relevant data and metadata in a secured environment between users having differing requirements it was necessary to have requisite networking and hardware infrastructure. Since GIS datasets can be very large and dense, moving even subsets of these datasets across the Enterprise network for manipulation or viewing purposes can have serious network performance implications. This capacity planning was done by the systems design team based on the complexity and size of spatial datasets, number of concurrent users and the spatial transactions to be done over the network (Peters, 2004). Detailed discussion of this design is beyond the scope of this paper. Since the spatial data repository is likely to get large and will be accessed by users all over, a suitable RDBMS selection was also necessary. Lastly, the client software for accessing the geodatabase was also a major requirement - keeping in view that all users may not have the same purpose.

Table 7.1 : User Requirements in GSI.

GSI perspective	Role	User Category	Requirement
Projects / Divisions	Generates data through fieldwork and / or laboratory work, often generate derived or updated datasets	Data owner /creator, data user, rarely data manager	Needs to load and/or update data to the central repository, Needs access to existing data and metadata relevant to the ongoing project work.
Geodata Division and Map & Cartography Division	Manages all spatial and non-spatial data,	Primarily Data Manager, also plays the role of data user	Infrastructure to ensure efficient and standardized means to manage and disseminate vast volume of data
Top Management	Responsible for institutional policy making, supervision and decision making	Data User	View and explore all data and metadata

7.4 System Design and Architecture

Based on the requirements discussed above, the following technical objectives guided the design and system development of the Enterprise GIS:

- Provide an organizational infrastructure for storing existing data
- Create an user interface to search for and view data
- Provide information to users regarding availability and usefulness of data
- Ensure data standardization guidelines are followed by the data owners / creators
- Ensure data security
- Integrate the existing GIS infrastructure into the fold of Enterprise GIS
- Utilise in-house expertise
- Eliminate challenges faced while using spatial data - duplicate datasets, different projection systems, different attribute structure, etc.

In other words GSI required GIS interoperability through Enterprise GIS (EGIS). Adoption of EGIS would put an end to individual project-wise, different datasets. There will be no duplication and redundancy, no data reformatting for other applications, no coordinate system transformations. EGIS lets organizations have all spatial data in one location, accessible via an Intranet or the Internet and stops degrading the accuracy and quality of the datasets. And that is now possible since spatial data can be stored inside a Relational Database (RDBMS). RDBMS offer efficient data retrieval for organizations with many users accessing the same data (Farley, 2004). Spatial data stored in a RDBMS can be made accessible over a network and up-to-date data can be made available to users throughout the enterprise. In this case data is edited directly on the server, and once data maintenance on the new version is finished, updated spatial data can be made available to users throughout the enterprise. On the contrary, data maintained on local systems must always be physically distributed to those needing access to updated GIS data.

7.4.1 Architecture Overview

EGIS applications and tools will be used at CHQ, RHQ, OU and Divisions. It will centre around a data warehouse, which will act as a single data source for all users spread across the organization. In principle, a data warehouse is a large database organizing operational data in a repository for easy query and analysis and supports decision-making. In the present case the data warehouse integrates the spatial component of the data and is a spatial data warehouse (SDW) (ESRI, 1998, Saha, 2004). The technical components of SDW include an Internet Map Server (IMS) and a Spatial Database Engine (SDE) (Fig 7.1). IMS technology is the result of convergence of GIS and Internet. IMS or Internet GIS tools provide mapmakers the possibility to publish their maps on the Web and offer their users a wide combination of display and query functionality. Through Internet GIS, users are able to access GIS applications without purchasing GIS software, but only by using a Web browser. Detailed maps can be generated from huge databases of spatial information and distributed over the Web (Saha, 2002). SDE facilitates storing and managing of spatial data along with attributes in a single physical database and provides a gateway for GIS to interact with RDBMS (Fig 7.1). GSI adopted the ESRI ArcGIS software and Oracle 10G RDBMS as the building blocks for implementing the Enterprise GIS (Farley, 2004). ESRI ArcSDE software (ESRI, 2005) was chosen as the SDE component and ESRI ArcIMS (ESRI, 2004a) as the Internet Map Server component. ArcSDE also serves spatial data to the ArcGIS 9.x Desktop products and through ArcIMS, and is the key component in managing the SDW.

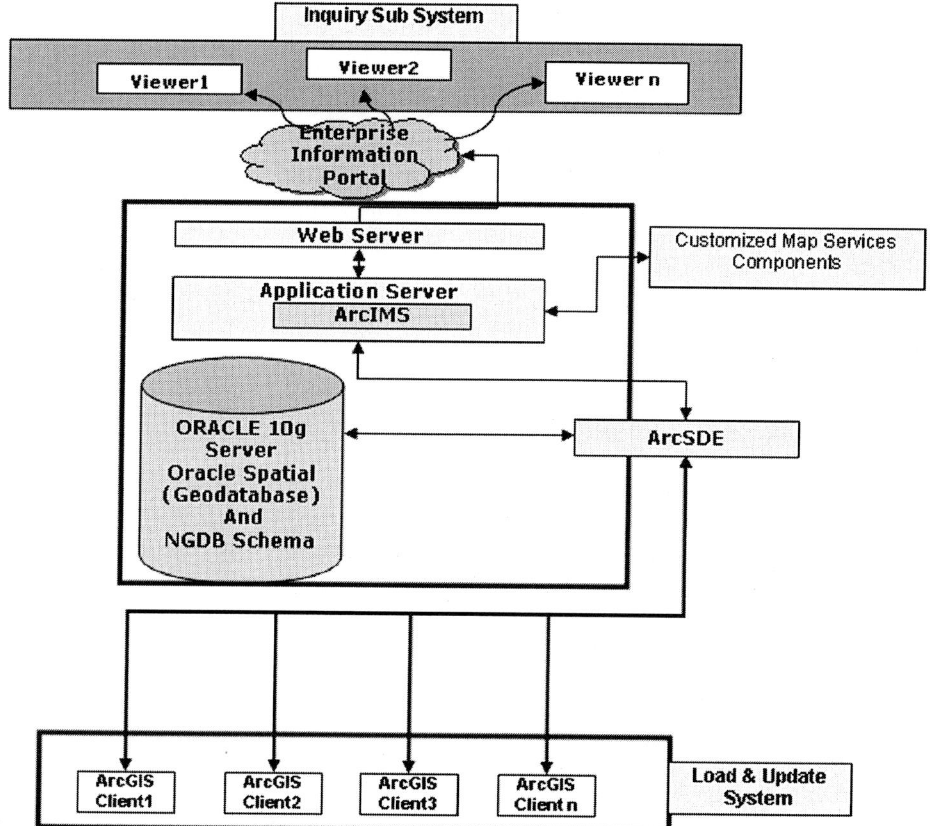

Fig. 7.1 : High level architecture of the Enterprise GIS in Geological Survey of India.

A data model was needed which can leverage the RDBMS capabilities to scale GIS datasets to extremely large sizes and to serve a number of users concurrently. The aim was to take advantage of advances in relational database technology in order to provide the enterprise features like security, multiple concurrent user access, and spatial indexes lacking in a file-based data model. The spatial index scheme on the spatial data allows a user to extract and render quickly a subset of a large spatial data layer. This capability permits to move away from the tile-based spatial data model and create seamless data layers. The Geodatabase model of ESRI is one such object-relational data model (URL 5). Within the Geodatabase, there are several base object classes to enable the storage and management of spatial objects. Feature Classes are the most basic type of Geodatabase object, whereas Feature Datasets are intended to store a group of feature classes that share common spatial relationship. In the present case Feature Datasets has been used as a mechanism for logical data organization within the SDW.

In a multi-user GIS environment, editing normally occur over a long period. During such 'long transactions' it is necessary to make data available to other users without duplicating the data and without applying prohibitively restrictive data locks. This is achieved by 'Versioning'. The ArcSDE/ Oracle approach of storing spatial data supports versioning (ESRI, 2004b). This implies that a default copy of the data remains intact in the database, while the data owner/creator makes certain changes to a different 'version' of the data. A user with access to that data would not see those updates until those changes were registered to the default copy by the data manager (Table 7.1). This concept of versioning has been adopted in the present case.

7.4.2 Functional Overview

The functional components of the Enterprise GIS as required by GSI can be classified as follows:

- Centralized multi-user geodatabase (Spatial Data Warehouse)
- Load and update sub system
- Metadata editor and explorer
- Inquiry sub system

The spatial data and its mandatory attributes will be stored in the multi-user Geodatabase in Oracle through ArcGIS clients, while its attribute data will be populated directly to NGDB schema using web forms. The Load and Update utility and the Metadata Editor will be incorporated with ArcGIS desktop clients, while the Inquiry system and the Metadata Explorer will be integrated with the EIP. The ArcGIS clients and ArcIMS service will connect to the Oracle RDBMS through ArcSDE service. Web browsers will display the map using ArcIMS and ArcSDE service and will show its corresponding attribute data from the NGDB schema residing in Oracle server. The data editing and updating through ArcGIS will be done using versioning.

7.4.3 The Spatial Data Warehouse

The ArcSDE multi-user geodatabase will act as the SDW for all users across the organization. The geoscientific domains present in the legacy NGDB schema has been organised in individual feature datasets and different layers within each domain as feature classes (Table 7.2). The data will be stored in the geodatabase using UTM coordinate system in meters, and the WGS84 geodetic system.

While designing the Geodatabase, multi-level development strategy was followed (Asch *et al.,* 2004). Conceptual design of geodatabase was developed after identification of domains and the associated spatial layers (Fig 7.2). For each of these domains the spatial schema was conceptualised and the conceptual design was translated into logical geodatabase entities (domain creation, feature datasets, feature classes and their attributes). Finally, the logical geodatabase schema led to the physical ArcSDE geodatabase design. This was accomplished by generating geodatabase schema directly from ArcCatalog. This approach ensures that changes in any level do not impact the level above it, i.e. changes in the physical level will not impact logical and conceptual designs (Asch *et al.,* 2004).

Table 7.2 : Partial Layer (Feature Class) structure of Map50K domain (Feature Dataset).

S.No	Layer Name (Feature Class)	Geometry Type	Description	Table in NGDB schema containing attribute data
1	LI	Polygon	Lithological Unit	M_LITHOLOGICAL_ UNIT
2	BO	Line	Lithocontacts	M_BOUNDARY
3	DY	Line	Dykes	M_DYKE
4	ZS	Polygon	Zonal Structures	M_ZONAL_ STRUCTURES
5	FA	Line	Faults	M_FAULT
6	FO	Line	Regional folds	M_FOLD
7	OS	Point	Oriented structural elements	M_ORIENTED_ POINT

The concept of versioning has been implemented to store, manage and retrieve spatial data in the multi-user geodatabase. GIS Administrator at CHQ will have full access to the different domains of the spatial datasets and will be responsible for creation and maintenance of users at RHQ's and Wings to provide access to the various logical versions of the geodatabase. The GIS Administrator will also create and assign the logical parent and child versions of the geodatabase to the concerned RHQ and Wings.

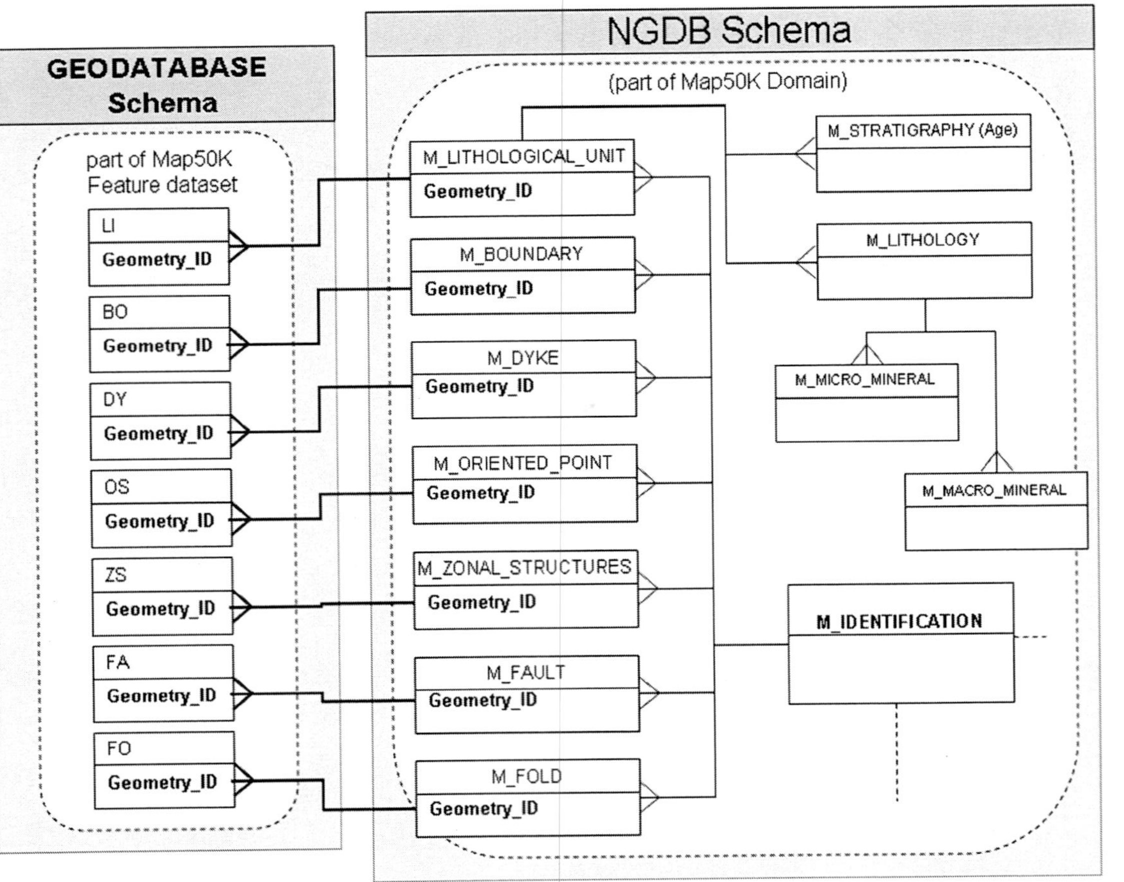

Fig. 7.2 : Part of Map50K feature dataset and corresponding Map50K domain in NGDB schema.

The workflow for versioning will initiate at the OU level. The OU GIS administrator after uploading data to the concerned logical version will send a notification to respective RHQ. RHQ administrator will create a child version of OU and reconcile & post the data to RHQ version after necessary verification (Fig 7.3). Before posting the spatial data to the main database, RHQ users will have the privilege to make certain changes in spatial data for resolving inter-Operational issues. RHQ will finally merge the spatial data from different OUs under its jurisdiction. A notification will be sent to CHQ to reconcile the final RHQ level spatial data to CHQ version. The CHQ administrator will validate inter-Regional conflicts prior to porting into the final production multi-user geodatabase. CHQ will have administrator rights for controlling the work at OU and RHQ level.

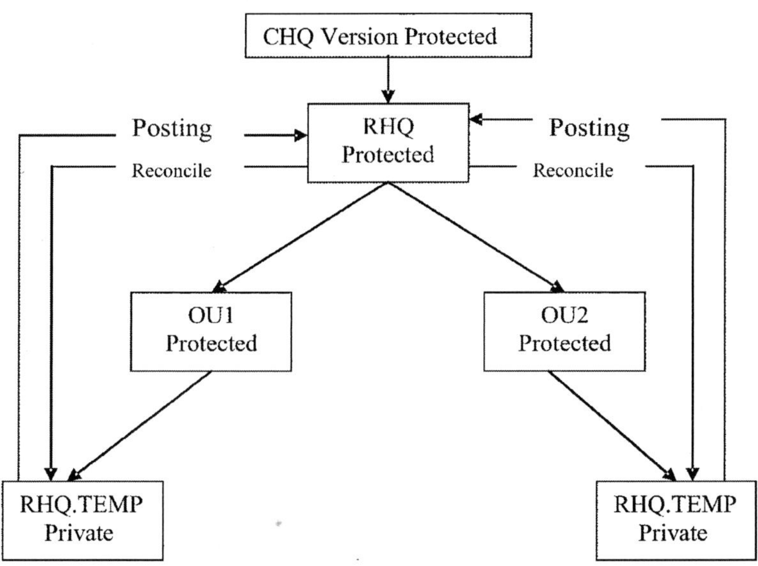

Fig. 7.3 : Versioning workflow.

Similar versioning workflow will be adopted for Coal and Marine Wing of GSI to manage and store the spatial data in the SDW.

7.4.4 Load and Update Subsystem

This utility has been configured on the ArcInfo desktops and is used to migrate the spatial datasets. This is in essence an extraction-transformation-loading tool, which extracts data from sources, transforms it and loads to the data warehouse (Meyerhanke, 2001). Data owners will be able to import the legacy and current data along with the mandatory attribute field data into concerned version of the multi-user geodatabase with the help this system. Presently GSI has large amount of spatial data in the form of ArcInfo Coverages, shapefiles, CAD (DWG/DXF) and 'X, Y' data. The corresponding attributes of geometry reside in the NGDB schema. Prior to migration of geometry into the geodatabase, the system will validate the layer name, geometry, projection status and the corresponding attribute information attached with the geometry. Spatial reference of legacy data could be in varying projection systems, which may or may not be compatible with each other. This issue has been resolved through the Load and Update Utility, which is capable of reprojecting the geospatial data into the SDW projection system. The attribute information attached with the geometry will act as composite primary key to establish the linkage between the geodatabase schema and NGDB schema. Spatial layers for geochemical, geophysical and drilling domain will be generated from the existing X and Y information in NGDB Schema. The necessary primary key field values will be fetched from the NGDB schema during the geometry creation process. Point geometry will be generated in this way to prevent duplication by creation through digitisation. The attribute data related to the new spatial data will be entered in the NGDB schema through Web-enabled forms through the EIP. Error log reports will be generated for the invalid geometries encountered during the process of migration.

7.4.5 Metadata Editor and Explorer

A Metadata Editor has been customised on top of ArcCatalog based on the NSDI Metadata structure (NSDI, 2003). The utility has been designed to capture the metadata from legacy spatial data. It can also be used for editing and exporting the metadata into XML format. The exported XML files will be stored into the document management system of the EIP. A customized XSL stylesheet has been included in the ArcCatalog stylesheet list for displaying the metadata in ArcCatalog and in the Metadata Explorer Web application. Users can search metadata for their area of interest through this application using keywords, attributes or geographical extent. The Metadata Explorer is actually a map based search interface powered by ArcIMS accessible to all users through the EIP. The search interface will display result both graphically and in tabular form. Simple tools for search and navigation in any cardinal direction have been provided in the interface.

7.4.6 The Inquiry Subsystem

This system implemented using ArcIMS will provide spatial data access to all authorised users over the network. Access to this web-based application is once again through the GSI EIP and any user having the need can visualize and explore maps by the means of a standard browser. The inquiry system interface will provide all standard navigational and spatial query tools. It will also allow thematic map generation based on any attribute. Legend will be displayed dynamically based on the features displayed. User will be able to compose and print a map of his choice after exploratory visualization. It will provide a synoptic view of information that may reveal data and knowledge gaps and thereby augment decision-making. Several customised map services will be made available to the users of GSI based on the user groups. The layer-based security will be implemented to restrict the access based on the user group. Authorised users can query even deeper levels of attribute data stored in the NGDB application schema through the Inquiry system.

7.4.7 Security Mechanism

The security needs of an organization change dramatically as soon as any group of users requires multiple concurrent use of any information asset. To ensure authorised use, the user credentials will be authenticated using Oracle Internet Directory (OID) service (URL 6). OID is a standard LDAP based directory that provides a single, centralized repository for all user data. It allows sites to manage user identities, roles, authorization and authentication credentials, as well as application specific preferences and profiles in a single repository. The Web-based Inquiry System is designed to allow the GIS administrator to set the type of access to data allowed for any given user. The administrator assign each spatial domain and layer a value in terms of what type of user is allowed to have access to it. Based on the authentication the links will be activated to enable the user to access the various map services.

For the users having access to ArcGIS desktops the security concern is different. The spatial data workflow in GSI allows each Region and Wing to edit their own data, but have read-only access to spatial data generated by other Regions. To maintain the same authorization level in the EGIS environment, it was necessary to integrate a security control, which ensures that every Region has write access to its data only. Since spatial data now resides in Oracle, it was possible to fulfil this requirement using row level security (RLS) mechanism or fine-grained access control (FGAC). FGAC essentially rests on setting an application role automatically when the user logs in, and then setting an appropriate SQL predicate based on the role (URL 7).

Accordingly, a SAG level officer, heading a Region will be able to view all data using the inquiry system, but will not have write access. A project level officer, will have access to all data relevant to his project, but will have write access to only his project data.

The EGIS is largely an Intranet portion of the GSI EIP and that is protected by the network infrastructure and requisite firewalls from the portion, which is open to the World Wide Web.

7.5 Discussion

The present EGIS is an effort to streamline geospatial data management in GSI. Globally, there has been an effort to create dynamic digital geological map data for the world (URL 8, Asch *et al.,* 2004) and this initiative can provide the Indian input to that effort.

Given the trend of analogue to digital conversion in GSI, lot more spatial data will become available within a few years. Therefore, the challenge would be to scale the system to acquire and serve more spatial data to the users. Network bandwidth is steadily improving and becoming accessible to more users – thus it will be possible to serve heavy data and applications over the Internet to a much larger community. Future work will thus focus on making the system more interoperable and publishing the data over the World Wide Web through OGC compliant web services (URL 9). It will also become necessary to prescribe digital cartographic standards for serving high-quality consistent maps (Saha *et al.,* 2003).

Above all, as major RDBMS and GIS technologies are converging, integration of spatial concepts into the rest of GSI's business applications will become a reality. GSI is already investing a fair amount of time and energy to achieve that milestone. Evidently, only a large, cooperative, well-coordinated and sustained effort will let this goal materialise.

7.6 Conclusions

Enterprise GIS can significantly improve the availability and consistency of spatial data holding of an organization. It will reduce inefficiencies in information access and exchange, and thus lead to cost-effective, efficient and strategic decision making. The primary objective in GSI is to make all geological map data accessible and in the course of next few years the goal would be to collate all data in the spatial data warehouse. In this way GSI will have all spatial data in a standardized format, though inter-Regional 'boundary faults' may still exist. In fact, by providing a synoptic view of Indian Geology, the Inquiry System will help easy identification of these problem areas. Creating seamless layers for the entire geographic extent of the country is the ultimate goal. To achieve that GSI would first identify the discrepancies and conflicts and resolve them gradually. There would be areas, which may need further mapping to resolve the conflicts, but in this case it will be much easier to pinpoint the problem area than ever before. This approach is considered pragmatic considering the complex, interpretive nature of geological information. Similar approach is being adopted for the digital map of the world through the 'OneGeology' movement (URL 8). GSI intends to start small and build incrementally. Accordingly, the Enterprise GIS will be rolled out to small numbers of users initially and will be tested for performance under as different conditions as possible.

Acknowledgements

The authors are deeply indebted to Shri S. R. Sengupta, the driving force behind this whole endeavour. They also express their heartfelt gratitude towards their co-workers, who were epitome of co-operation. Finally, thanks are due to the Enterprise GIS application development team of TCS for their dedicated participation.

Bibliography

Asch K., Brodaric B., Laxton J. L. and Robida F. (2004). An international initiative for data harmonization in geology, Proceedings: 10th EC GI & GIS Workshop, ESDI State of the Art, Warsaw, Poland.

Brodaric B. (2004). The design of GSC Field*Log*: ontology-based software for computer aided geological field mapping, *Computers & Geosciences*, 30 pp 5-20

Croswell (2000). The role of standards in support of GDI, in R. Groot and J. McLaughlin (eds.) Geospatial Data Infrastructure: Concepts, Cases and Good Practice, Oxford University Press, pp 57-83

ESRI (2005). Understanding ArcSDE, ESRI documentation (available at www.esri.com)

ESRI (2004a). ArcIMS 9 Architecture and Functionality. ESRI white paper (available at www.esri.com)

ESRI (2004b). Versioning. ESRI Technical paper (available at www.esri.com)

ESRI (1998). Spatial data warehousing, ESRI white paper (available at www.esri.com)

Farley, (2004). Oracle 10g: A location-enabled platform for enterprise GIS and Core Business Applications, *Earth Observation Magazine*, www.eomonline.com

GSI (2006). Annual Programme of GSI: Field Season 2006-07. GSI, Kolkata

GSI (1997). Setting up of Geoscientific Data Centre at GSI: Final report. Unpublished Report of GSI. GSI, Kolkata

Meyerhanke H. (2000). Software engineering aspects of Spatial Data Warehousing, <http://www.meyerhenke.de/cs/inhalt.htm> (last accessed 22. 08.07)

NSDI (2003). NSDI Metadata standard version 3.0, ISRO, DOS, Govt. of India

Peters D., (2004). System Design Strategies: An ESRI technical Reference document, ESRI, California (last updated in July 2007, available at www.esri.com)

Saha A., (2004). Spatial Data Warehousing for GIS-enabled Enterprise Information Portal at GSI, Proceedings: Workshop on Digital Techniques in Geosciences, Project INDIGEO, GSI, Hyderabad

Saha A., Ganguly D., Sengupta S.R., Mukhopadhyay B., Hazra N., (2003). Digital Cartographic Standards for 1:50,000 geologic Map: some important considerations, Proceedings: NSDI Seminar, MapIndia 2003 International Conference, New Delhi.

Saha A., Ganguly D., (2003). Geological Mapping in India, *Kartografisch Tijdschrift*, Nederlandse Vereniging voor Kartografie, XXIX, pp. 20-28

Saha A. (2002). Maps for geology on the Internet: study area-Tabernas Basin, South Spain, Unpublished Report, PM degree course in Geoinformatics, Enschede: ITC.

Saha A., Mukhopadhyay B., Rajendran N. (2001). GIS integration of Exploration Datasets to Produce a Gold Potential Map for Gadag Schist Belt, Karnataka, India, *Indian Mineral*, 55, pp. 171-184

URL 1 GSI <http://www.gsi.gov.in> (last accessed on 24.08.07)

URL 2 ESRI: Extending GIS to Enterprise Applications <www.esri.com/library/whitepapers/pdfs/idc_enterprise_apps_feb_2005.pdf> (last accessed on 22.08.07)

URL 3 NeGP <http://www.mit.gov.in/> (last accessed on 24.08.07)

URL 4 NSDI <http://gisserver.nic.in/nsdiportal/> (last accessed on 24.08.07)

URL 5 ArcGIS Desktop Help: <http://webhelp.esri.com/arcgisdesktop/9.2/> (last accessed on 23.08.07)

URL 6 Oracle Internet Directory <http://www.oracle.com/technology/products/oid/index.html > (last accessed on 29.08.07)

URL 7 Oracle Virtual Private Database policy tips: <http://www.dba-oracle.com/art_builder_vpd.htm> (last accessed on 26.08.07)

URL 8 OneGeology <http://www.onegeology.org/> (last accessed on 24.08.07)

URL 9 OGC <http://www.opengeospatial.org/> (last accessed on 19.08.07)

Witkowski, M. S., Rich, P. M., Keating, G. N., Linger, S. P. and Riggs T. L. (2002). Spatial data warehouse design for Enterprise GIS, *GI Science*, pp 283-286.

Annexure

List of Abbreviations Used

Abbreviation/ Acronym	Description
AMD	Atomic Minerals Division
ArcGIS	ESRI GIS product family
ArcINFO	ESRI GIS desktop product
BRGM	Bureau de Recherches' Geologique et Minerals, France
CGWB	Central Ground Water Board
CHQ	Central Head Quarter
DGM	Department of Geology and Mines
DOS	Department of Space
EGIS	Enterprise GIS
EIP	Enterprise Information Portal
ESRI	Environmental System Research Institute
GDM	A Geological Data Management tool (by BRGM)
GIS	Geographical Information System
GSI	Geological Survey of India
ISRO	Indian Space Research Organisation
JAG / SAG	Junior Administrative Grade / Senior Administrative Grade
JTS / STS	Junior Time Scale / Senior Time Scale
LDAP	Lightweight Directory Access Protocol
MECL	Mineral Exploration Corporation Limited
NGDB	National Geoscientific Database
NSDI	National Spatial Data Infrastructure
OGC	Open Geospatial Consortium
OID	Oracle Internet Directory
OU	Operational Unit
RHQ	Regional Head Quarter. Generally refers to Wings also.
SDW	Spatial Data Warehouse
SynARC	An extension for ArcGIS (by BRGM)
XML / XSL	Extensible Mark-up Language/ Extensible Style Sheet Language

Index

A

B

C

T

U

V

Z